SpringerBriefs in Earth Sciences

SpringerBriefs in Earth Sciences present concise summaries of cutting-edge research and practical applications in all research areas across earth sciences. It publishes peer-reviewed monographs under the editorial supervision of an international advisory board with the aim to publish 8 to 12 weeks after acceptance. Featuring compact volumes of 50 to 125 pages (approx. 20,000–70,000 words), the series covers a range of content from professional to academic such as:

- timely reports of state-of-the art analytical techniques
- bridges between new research results
- snapshots of hot and/or emerging topics
- literature reviews
- in-depth case studies

Briefs will be published as part of Springer's eBook collection, with millions of users worldwide. In addition, Briefs will be available for individual print and electronic purchase. Briefs are characterized by fast, global electronic dissemination, standard publishing contracts, easy-to-use manuscript preparation and formatting guidelines, and expedited production schedules.

Both solicited and unsolicited manuscripts are considered for publication in this series.

More information about this series at http://www.springer.com/series/8897

Alexey Nekrasov

Foundations for Innovative Application of Airborne Radars

Functionality Enhancement for Measuring the Water Surface Backscattering Signature and Wind

Second Edition

 Springer

Alexey Nekrasov
Institute for Computer Technologies
and Information Security
Southern Federal University
Taganrog, Russia

Technical University of Košice
Košice, Slovakia

ISSN 2191-5369   ISSN 2191-5377   (electronic)
SpringerBriefs in Earth Sciences
ISBN 978-3-030-62941-0   ISBN 978-3-030-62942-7   (eBook)
https://doi.org/10.1007/978-3-030-62942-7

This Springer imprint is published by the registered company Springer Nature Switzerland AG
The registered company address is: Gewerbestrasse 11, 6330 Cham, Switzerland

*To Professor Dr. Sc. Ashot Garnakeryan,
my scientific supervisor*

Acknowledgements

I would like to express my sincere thanks to Prof. Dr. Colin Fidge and Prof. Dr. David Lovell (The Queensland University of Technology), Prof. Dr. Jason H. Middleton (The University of New South Wales), Prof. Dr. Joel T. Johnson (The Ohio State University), Prof. a.D. Dr.-Ing. Dr. h.c. Klaus Schünemann, Prof. a.D. Dr.-Ing. Arne Jacob and Prof. Dr.-Ing. Udo Carl (Hamburg University of Technology), Prof. Dr.-Ing. Reinhard Knöchel (Christian-Albrechts-University of Kiel), Dr. Wolfgang Rosenthal (GKSS Research Center, Geesthacht), Prof. Dr. hab. Adam Krężel (University of Gdańsk), Prof. Dr. J. Pereira Osório (University of Porto), Prof. Dr. ir. Peter Hoogeboom and Prof. Dr. ir. Leo P. Ligthart (Delft University of Technology), Prof. Dr. Maurizio Migliaccio, Prof. Dr. Paolo Corona and Prof. Dr. Renato Passaro (University of Naples "Parthenope"), Prof. Dr. h.c. Ing. Stanislav Kmeť, Assoc. Prof. Dr. Mária Gamcová, Assoc. Prof. Dr. Pavol Kurdel, Prof. Dr. Ján Labun and Assoc. Prof. RNDr. František Olejník (Technical University of Košice) for their research opportunity provided, and to the Ministry of Education and Science of Russia, the Endeavour Awards Program, Australia, the Fulbright Program, USA, the German Academic Exchange Service (DAAD), the Information Processing Center (OPI), Poland, the Institute for International Scientific and Technological Co-operation (ICCTI), Portugal, the Netherlands Organization for Scientific Research (NWO), the National Research Council of Italy (CNR), the Slovak Academic Information Agency (SAIA) for their research grants and fellowships.

Contents

About the Author

Alexey Nekrasov received the diploma in radio engineering, the Ph.D. degree (Candidate of Technical Science) and the academic title of Associate Professor from Taganrog State University of Radio Engineering (TSURE), Taganrog, Russia, in 1991, 1998, and 2000, respectively.

From 1991 to 1999, he was an Assistant Professor at TSURE. Since 1999, he has been an Associate Professor at the Southern Federal University (formerly TSURE), Russia. Since 2016, he has also been a Senior Scientist at Saint Petersburg Electrotechnical University, Russia, and he is also with the Technical University of Košice, Slovakia.

He has been an expert of the Russian Academy of Sciences, the Russian Science Foundation, the Fulbright Program in Russia, a recipient of 24 international grants and awards, and a reviewer for IEEE Transactions on Geoscience and Remote Sensing, IEEE Geoscience and Remote Sensing Letters, IEEE Antennas and Wireless Propagation Letters, Remote Sensing, Atmosphere, Climate, Symmetry, Journal of Marine Science and Engineering, IET Radar, Sonar & Navigation, IET Image Processing, and ISPRS International Archives of the Photogrammetry, Remote Sensing and Spatial Information Sciences.

His research interests include sea and land remote sensing from aircraft and satellites, oceanography, sea winds, scattering from rough surfaces, radar system design, and information security.

Acronyms

ARA   Airborne Radar Altimeter
APR   Airborne Precipitation Radar
AWR   Airborne Weather Radar
DNS   Doppler Navigation System
FM-CW   Frequency-modulated continuous-wave
GPS   Global Positioning System
NRCS   Normalized radar cross section
PC   Personal computer
RMS   Root mean square
SAR   Synthetic Aperture Radar

Symbols

A	Zeroth Fourier term
a	Antenna length in direction of flight
a_0, a_1, a_2	Coefficients
B	First Fourier term
b	Antenna width
C	Second Fourier term
c	Speed of light
F_D	Doppler frequency
$F_{D1.a}, F_{D2.a}$	Frequency limits for aft-Doppler filter
$F_{D1.f}, F_{D2.f}$	Frequency limits for fore-Doppler filter
g	Acceleration of gravity
$g(\theta', \phi')$	Antenna pattern shape
$g(\theta')$	Pattern factor in across-track direction
$g(\phi')$	Pattern factor in along-track direction
G	Antenna gain
G_1, G_2	Parameters
H	Altitude of flight
h	Coefficient
h_c	Cross-wind coefficient
h_d	Down-wind coefficient
h_u	Up-wind coefficient
h_w	Water waves height
k_1, k_2	Coefficients
K_w	Coefficient taking into account an expansion of the antenna beamwidth
$K_{w.s}$	Simplified coefficient K_w
L_s	System losses
N	Number of sectors observed
N_s	Number of samples that can be obtained from a sector observed
P_r	Power received from a resolution cell

$\bar{P}_r$	Average power received
P_t	Transmitted power
Q	Coefficient
Q_c	Cross-wind coefficient
Q_d	Down-wind coefficient
Q_u	Up-wind coefficient
$R(0°)$	Fresnel reflection coefficient at normal incidence
R_g	Ground range
$R_{t.c}$	Radius of turn for the selected cell middle point
$R_{t.fa}$	Radius of an aircraft turn
$S^2(\alpha)$	Water surface slope variance in the azimuthal direction α
S_c	Cross-wind standard deviation of slopes
S_{min}	The least value of summation results
S_u	Up-wind standard deviation of slopes
t	Time
$T_{360°}$	Time of 360-degree turn
T_s	Time of a sector view
V	Speed of flight
V_g	Aircraft velocity relative to the ground
U	Wind speed
U_{10}	Wind speed at 10 m above the water surface
$U_{12.5}$	Wind speed at 12.5 m above the water surface
$U_{19.5}$	Wind speed at 19.5 m above the water surface
U_z	Wind speed at anemometer height of $z$meters
α	Azimuth illumination angle relative to up-wind direction; azimuth angle between up-wind direction and an aircraft course; azimuth angle between up-wind direction and first NRCS azimuth
α_{1an}	Up-wind direction obtained from the first annulus cells
α_{2an}	Up-wind direction obtained from the second annulus cells
α_q	Angle of wind in a quadrant
α_s	Middle azimuth of a sector
$\beta_{dr.max}$	Maximum possible drift angle
Γ_0	Mounting angle for a beam axis in the horizontal plane
$\gamma_1, \gamma_2, \gamma_3$	Coefficients
γ_{fa}	Roll angle of an aircraft (flying apparatus)
$\Delta\alpha_b$	Angular resolutions in azimuthal plane
Δf_D	Width of a Doppler spectrum of a reflected signal
$\delta K_{w.s}$	Relative error for simplification of the coefficient K_w
Δr_{az}	Azimuth resolution
Δr_{gr}	Ground-range resolution
Δt	Time passed after receiving the last energy from nadir point
$\Delta\alpha$	Azimuth angular size; angular resolutions in azimuthal plane
$\Delta\alpha_1$	Azimuth angular sizes of cells of first annulus
$\Delta\alpha_2$	Azimuth angular sizes of cells of second annulus

$\Delta\alpha_s$	Azimuth size of a sector
$\Delta\alpha_w$	Width of a wide azimuth sector
$\Delta\theta$	Angular resolution in vertical plane
$\Delta\theta_1$	Angular incidence widths for first annulus zone
$\Delta\theta_2$	Angular incidence widths for second annulus zone
$\Delta\sigma_2^\circ$	NRCS difference obtained with fore-Doppler and aft-Doppler filters for a two-cell geometry
$\Delta\sigma_{2.\max}^\circ$	Maximum possible NRCS difference obtained with fore-Doppler and aft-Doppler filters for a two-cell geometry
$\Delta\sigma_4^\circ$	NRCS difference obtained with the fore-Doppler and aft-Doppler filters for a four-cell geometry
$\Delta\sigma_{4.\max}^\circ$	Maximum possible NRCS difference obtained with fore-Doppler and aft-Doppler filters for a four-cell geometry
η_0	Mounting angle for a beam axis in inclined plane
θ	Incidence angle
θ_1	Incidence angle for first annulus zone
θ_2	Incidence angle for second annulus zone
θ'	Incidence angle relative to antenna beam axis
θ_0, θ_m	Incidence mounting angle for an antenna beam axis
θ_a	Antenna beamwidth
$\theta_{a.h}$	Antenna beamwidth in horizontal plane
$\theta_{a.v}$	Antenna beamwidth in vertical plane
$\theta_{a.incl}$	Effective antenna beamwidth in inclined plane
$\theta_{az.ef}$	Effective azimuth beamwidth
θ_e	Incidence angle corresponding to angular location of annulus external border
$\theta_{el.ef}$	Effective antenna elevation beamwidth
θ_{fa}	Pitch angle of an aircraft (flying apparatus)
θ_i	Incidence angle corresponding to the angular location of annulus internal border
θ_p	Angle for a pulse-limited footprint
$\pm\theta_s$	Cross-track elevation range
λ	Carrier wavelength
σ_{cell}	Radar cross section of a selected cell
σ°	Normalized radar cross section
$\sigma_{360^\circ}^\circ$	360° azimuthally averaged NRCS
σ_{an}°	Azimuthally integrated NRCS obtained from annulus zone
σ_c°	Cross-wind NRCS
σ_d°	Down-wind NRCS
σ_{ad}°	NRCS obtained with aft-Doppler filter
σ_{cal}°	Calculated NRCS value
σ_{fd}°	NRCS obtained with fore-Doppler filter
$\sigma_{\max}^\circ$	Main maximum value for azimuth NRCS set
σ_u°	Up-wind NRCS

$\sigma_w^\circ(0^\circ)$	NRCS measured by a scatterometer having a nadir-looking wide-beam antenna
σ_γ^2	Averaged variance of the sea-surface slopes
τ	Pulse duration
ϕ'	Azimuth angle relative to the antenna beam axis
φ	Current horizontal angle of the selected sell relative to the aircraft course
ψ	Aircraft course
ψ_b	Azimuth direction of an antenna beam relative to an aircraft course corresponding to a real observation azimuth angle of a sector
ψ_c	Angle between an aircraft course and the azimuth of the first Doppler cell
ψ_d	Angle between the aircraft course and the azimuth of the cell selected from the second annulus
ψ_s	Real observation azimuth angle of a sector
$\psi_{s.b}$	Real observation azimuth angle of sector beginning
$\psi_{s.e}$	Real observation azimuth angle of sector end
ψ_w	Navigational direction of wind
$\psi_{\sigma_{\max}^\circ}$	Azimuth of the principal maximum of NRCS curve
ψ_{ψ_s}	Aircraft course corresponding to real observation azimuth angle of a sector

Chapter 1
Introduction

Water covers over 70 % of our planet's surface. The oceans interact with atmosphere to control and regulate environment on the Earth. Fed by the Sun, the interaction of land, ocean, and atmosphere produces the phenomena of weather and climate. Information on surface wind vector and wave height is assimilated into regional and global numerical weather and wave models, thereby extending and improving our ability to predict future weather patterns and sea/ocean surface conditions (Long et al. 1996).

During the last 50–60 years, meteorologists have begun to understand weather patterns well enough to produce sufficiently accurate forecasts of future weather patterns. At first, only local oceanic weather conditions were available from sparsely arrayed weather stations, ships, and ocean buoys. Later on, satellite and airborne remote sensing has improved the situation significantly. Satellite remote sensing has demonstrated its potential to provide measurements of weather conditions on a global scale as well as airborne remote sensing on a local scale.

Wind and wave measuring systems can be divided into two following categories (Komen et al. 1994):

(1) operational systems providing continuous, global, or regional observations for forecasting, data assimilation, and model validation;
(2) systems operating with higher temporal and spatial resolution during limited measurement campaigns in order to study the physical processes of wave generation and air/sea interaction.

Near-surface wind measurements over sea are very important for operational oceanography, as well as for meteorology and navigation.

On a global scale, the information about sea waves and winds, in general, can be obtained from a satellite using active microwave instruments: scatterometer, Synthetic Aperture Radar (SAR), and Radar Altimeter (Komen et al. 1994). However, for local numerical weather and wave models as well as for a pilot on an amphibious aircraft or seaplane who has to make a correct landing decision, local data on wave height, wind speed, and direction are required.

© The Author(s), under exclusive license to Springer Nature Switzerland AG 2021
A. Nekrasov, *Foundations for Innovative Application of Airborne Radars*,
SpringerBriefs in Earth Sciences, https://doi.org/10.1007/978-3-030-62942-7_1

Wind and wave measurements by those remote sensing instruments are based on features of microwave backscattering from the water surface. Scattering from a sea surface has been studied since the end of the 1940s (Moore and Jones 2004), and by now many wind-wave tank, platform, airborne, and satellite experiments have been performed to investigate wind and wave dependence of the backscattered signal parameters so as to use it as a tool for wind and wave remote retrieval.

A typical method for describing sea clutter is in a form of normalized radar cross section (NRCS), statistical distribution of the NRCS, amplitude correlation, and spectral shape of the Doppler returns (Money et al. 1997; Skolnik 2008). To study microwave backscattering signature of the water surface from aircraft, an airborne scatterometer is used. The measurements are typically performed at either a circle track flight using a fixed fan-beam antenna or a rectilinear track flight using a rotating antenna (Carswell et al. 1994; Li et al. 2011; Masuko et al. 1986; Wismann 1989). For such an airborne measurement of winds, antennas with comparatively narrow beams (beamwidth of $4°$ – $10°$) are commonly used. Unfortunately, a microwave narrow-beam antenna has considerable size at Ku-, X-, and C-bands that makes its placing on an aircraft difficult, especially on a seaplane or an amphibious aircraft. Therefore, a better way needs to be found.

At least two options can be proposed. The first option is to apply airborne scatterometers with wide-beam antennas as it can lead to the reduction of the antenna size. The second option is to use modified conventional navigation instruments of the aircraft in a scatterometer mode, which seems more preferable.

From that point of view, this monograph considers possibilities of enhancement of some modern airborne instruments in order to apply them for measuring water-surface backscattering signature and for estimating wind speed and direction over water, in addition to their standard application.

Airborne Frequency-Modulated Continuous Wave (FM-CW) Demonstrator System, Doppler Navigation System (DNS), Airborne Weather Radar (AWR), Airborne Radar Altimeter (ARA), and Airborne Precipitation Radar (APR) operating in a scatterometer mode are discussed with the possible measuring algorithms and recommendations on how to perform the measurements.

References

Carswell JR, Carson SC, McIntosh RE, Li FK, Neumann, G, McLaughlin DJ, Wilkerson JC, Black PG, Nghiem SV (1994) Airborne scatterometers: investigating ocean backscatter under low- and high-wind conditions. Proc IEEE 82(12):1835–1860

Komen GJ, Cavaleri L, Donelan M, Hasselmann K, Hasselmann S, Janssen PAEM (1994) Dynamics and modelling of ocean waves. Cambridge University Press, Cambridge, p 532

Li L, Heymsfield G, Carswell J, Schaubert D, McLinden M, Vega M, Perrine M (2011) Development of the NASA high-altitude imaging wind and rain airborne profiler. Proceedings of aerospace conference, Big Sky, MT, USA, 5–12 Mar 2011, pp 1–8

Long DG, Donelan MA, Freilich MH, Graber HC, Masuko H, Pierson WJ, Plant WJ, Weissman D, Wentz F (1996) Current progress in Ku-band model functions. Brigham Young University, USA, Tech Rep MERS 96-002, p 88

Masuko H, Okamoto K, Shimada M, Niwa S (1986) Measurement of microwave backscattering signatures of the ocean surface using X band and Ka band airborne scatterometers. J Geophys Res 91(C11):13065–13083

Money DG, Mabogunje A, Webb D, Hooker M (1997) Sea clutter power spectral lineshape measurements. Proceedings of Radar'97, Edinburgh, UK, 14–16 Oct 1997, pp 85–89

Moore RK, Jones WL (2004) Satellite scatterometer wind vector measurements—the legacy of the seasat satellite scatterometer. IEEE Geosci Remote Sens Newslett 132:18–32

Skolnik M (2008) Radar handbook. McGraw-Hill, New York, p 1330

Wismann V (1989) Messung der Windgeschwindigkeit über dem Meer mit einem flugzeugge-tragenen 5.3 GHz Scatterometer. Dissertation zur Erlangung des Grades eines Doktors der Naturwissenschaften, Universität Bremen, Bremen, Germany, S 119

Chapter 2
Water-Surface Backscattering and Wind Retrieval

Many researchers have been investigating the microwave backscattering signatures of water surface and solving the problem of remote measuring of wind speed and direction over water (Bentamy et al. 2012; Carswell et al. 1994; Chelton and McCabe 1985; Feindt et al. 1986; Giovanangeli et al. 1991; Hildebrand 1994; Long 2001; Masuko et al. 1986; Melnik 1980; Moore and Fung 1979; Nielsen and Long 2009; Plant 2003; Ward et al. 2008; Wismann 1989). However, the mechanics of interactions between water surfaces and microwaves has not been well studied in detail.

As it was mentioned earlier, a typical method for describing sea clutter is in the form of NRCS, statistical distribution of the NRCS, amplitude correlation and spectral shape of the Doppler returns (Money et al. 1997).

For description of radar backscatter from water surfaces, three major scattering models are used: Kirchhoff or physical optics model, composite-surface or two-scale model, and the Bragg model. The Kirchhoff model assumes a perfectly conducting surface (unless it is modified to include the Fresnel reflection coefficient) and applies from small to intermediate incidence angles without shadowing effects. Apart from the implicit dependence on the Fresnel coefficient, there is no polarization dependence. The two-scale model assumes that radar backscatter arises from a large number of slightly rough ripples, distributed over the long ocean waves. It has polarization dependence. These two models are generally used to interpret the data acquired by a synthetic aperture radar and real-aperture radar of a variety of sea/oceanic features, including swell waves and internal waves. The Bragg model applies only to slightly rough surfaces under low wind conditions (it is often used to describe the scattering from ripples in the two-scale model). The Bragg model has been used to interpret the ocean currents by high-frequency Doppler-radar measurements at large incidence angles (Ouchi 2000).

To explain adequately the microwave scattering signature of the water surface and to apply its features to remote sensing (including wind and rain measurements over sea (Nielsen and Long 2009), oil spills detection (Robbe and Hengstermann 2006), sea ice monitoring (Johannessen et al. 2007), a set of experiments, namely, experimental verification of the combined frequency, azimuth and incidence angles,

© The Author(s), under exclusive license to Springer Nature Switzerland AG 2021
A. Nekrasov, *Foundations for Innovative Application of Airborne Radars*,
SpringerBriefs in Earth Sciences, https://doi.org/10.1007/978-3-030-62942-7_2

and wind speed variations of the NRCS are required (Masuko et al. 1986). For that study, a scatterometer, radar designed for measuring the surface scatter characteristics, is used.

Research on microwave backscatter by the water surface has shown that the use of a scatterometer also allows an estimation of the near-surface wind vector because the NRCS of the water surface depends on wind speed and direction. Based on experimental data and scattering theory, a significant number of empirical and theoretical backscatter models and algorithms for estimation of near-surface wind vector over water from satellite and airplane have been developed (Fernandez et al. 2006; Isoguchi and Shimada 2009; Long et al. 1996). The accuracy of wind direction measurement by scatterometer is ±20°, and accuracy of wind speed scatterometer measurement is ±2 m/s in wind speed range 3–24 m/s.

SAR provides an image of roughness distribution on a sea surface with a large dynamic range, high accuracy, and high resolution. Retrieval of wind information from SAR images provides a useful complement to support traditional wind observations (Du et al. 2002). Wind direction estimation amounts to measuring the orientation of boundary-layer rolls in a SAR image, which are often visible as image streaks. Sea-surface wind direction (to within a 180° direction ambiguity) is assumed to lie essentially parallel to the roll or image-streak orientation. Wind speed estimation from SAR images is usually based on the scatterometer wind retrieval models. This approach requires a well-calibrated SAR image. Wind direction estimated from European remote sensing satellite SAR images is within a root mean square (RMS) error of ±19° of in situ observations, which in turn results in an RMS wind speed error of ±1.2 m/s in offshore measurements (Hasager et al. 2011; Wackerman et al. 1996) and from ±2 to ±3 m/s for coastal waters (Takeyama et al. 2012).

The radar altimeter also provides information on sea wind speed, which can be determined from the intensity of the backscattered return pulse, and on sea wave height, which can be deduced from the return pulse shape (Karaev and Kanevsky 1999). At moderate winds (3–12 m/s), the wind speed can be measured by an altimeter with an accuracy of about ±2 m/s. Typical accuracy of radar altimeter measurements of a significant wave height is on the order of ±0.5 m (or 10 %, whichever is higher) for wave heights between 1 and 20 m (Chelton et al. 2001; Komen et al. 1994). Unfortunately, altimeter wind measurements yield wind velocity magnitude only and do not provide information on wind direction (Chelton et al. 2001).

The backscatter of radio waves from sea surface varies considerably according to the incidence angle (Hildebrand 1994). Near nadir, there is a region of quasi-specular return with a maximum of NRCS that decreases when increasing the angle of incidence. Between incidence angles of about 20° and 70°, NRCS falls smoothly in a so-called plateau region. For medium incidence angles, microwave radar backscatter is predominantly due to the presence of capillar-gravity wavelets, which are superimposed on large gravity waves on the sea surface. Small-scale sea waves approximately one half the radar wavelength are in Bragg resonance with an incident electromagnetic wave. At incidence angles greater than about 70°, there is a "shadow" region in which the NRCS falls dramatically, due to the shadowing effect of waves closer to the radar blocking waves further away.

The wind blowing over the sea modifies surface backscatter properties. These depend on wind speed and direction. Wind speed U can be determined by a scatterometer because a stronger wind will produce a larger NRCS $\sigma°(U, \theta, \alpha)$ at a medium incidence angle θ, and a smaller NRCS at a small (near nadir) incidence angle. Wind direction can also be inferred because NRCS varies as a function of azimuth illumination angle α relative to the up-wind direction (Spencer and Graf 1997).

To retrieve the wind vector from NRCS measurements, a relationship between NRCS and near-surface wind, called the "geophysical model function," must be known. Scatterometer experiments have shown that the NRCS model function for medium incidence angles at appropriate transmit and receive polarization (vertical or horizontal) can be presented in several various forms. One of the widely used forms for the geophysical model function is as follows (Moore and Jones 2004; Spencer and Graf 1997)

$$\sigma_{pp}^{\circ}(U, \theta, \alpha) = A(U, \theta) + B(U, \theta)\cos\alpha + C(U, \theta)\cos(2\alpha), \qquad (2.1)$$

where subscripts pp represent the transmit and receive polarization (V – vertical, H – horizontal); $A(U, \theta)$, $B(U, \theta)$, and $C(U, \theta)$ are the Fourier terms that depend on sea-surface wind speed and incidence angle,

$$A(U, \theta) = a_0(\theta)U^{\gamma_0(\theta)}, \qquad (2.2)$$

$$B(U, \theta) = a_1(\theta)U^{\gamma_1(\theta)}, \qquad (2.3)$$

$$C(U, \theta) = a_2(\theta)U^{\gamma_2(\theta)}; \qquad (2.4)$$

$a_0(\theta)$, $a_1(\theta)$, $a_2(\theta)$, $\gamma_0(\theta)$, $\gamma_1(\theta)$, and $\gamma_2(\theta)$ are the coefficients dependent on the incidence angle. The term $A(U, \theta)$ equals the azimuthally averaged NRCS; the term $B(U, \theta)$ embodies the up-wind–down-wind asymmetry; and the term $C(U, \theta)$ represents the up-wind–cross-wind anisotropy. Detailed procedures to obtain the above terms from backscatter values in the up-wind, down-wind, and cross-wind directions have been reported in (Nghiem et al. 1997). The coefficients $a_0(\theta)$, $a_1(\theta)$, $a_2(\theta)$, $\gamma_0(\theta)$, $\gamma_1(\theta)$, and $\gamma_2(\theta)$ are determined empirically for an appropriate geophysical model function using airborne/satellite measurements and validating surface measurements. The coefficients are tabulated or presented as functions of incidence angle (Bentamy et al. 2012; Hans 1987; Long et al. 1996; Moore and Fung 1979; Wismann 1989).

From (2.1), we can see that an NRCS azimuth curve has two maxima and two minima. The principal maximum is located in the up-wind direction, the second maximum corresponds to the down-wind direction, and the two minima are in the cross-wind directions displaced slightly to the second maximum direction. With increase of the incidence angle, the difference between the two maxima and the difference between the maxima and minima become significant (especially at medium incidence angles); thus, this feature can be used for retrieval of wind direction over water (Ulaby et al. 1982).

Generally, the problem of estimating the navigational direction of the sea-surface wind ψ_w consists of defining the principal maximum of a curve of the reflected signal intensity (azimuth of the principal maximum of the NRCS curve $\psi_{\sigma^\circ_{max}}$)

$$\psi_w = \psi_{\sigma^\circ_{max}} \pm 180^\circ, \tag{2.5}$$

where $\pm$ before 180° means that if $0^\circ \leq \psi_{\sigma^\circ_{max}} < 180^\circ$, it should be converted into the wind navigational direction by adding 180° but if $180^\circ \leq \psi_{\sigma^\circ_{max}} < 360^\circ$ it should be converted by subtracting 180° as an angular value should always be in the interval $[0^\circ, 360^\circ[$.

The problem of deriving sea-surface wind speed consists of determination of the reflected signal intensity value from the up-wind direction or from some or all of the azimuth directions.

The azimuth NRCS curve can be obtained by stepping a narrow-beam antenna of a scatterometer through a range of angles over the water surface under the wind-wave tank and platform measurements (Giovanangeli et al. 1991; Snoeijl et al. 1992). At airborne measurements, the azimuth NRCS curve is obtained using a circle track flight for a scatterometer with an inclined one-beam fixed-position antenna or a rectilinear track flight for a scatterometer with a rotating antenna (Carswell et al. 1994; Li et al. 2011; Masuko et al. 1986; Wismann 1989).

Also, the wind speed over water can be measured by a scatterometer with a nadir-looking antenna (altimeter). The measurement is based on specular returns from the water surface, and for the wind speed estimation, a wind retrieval algorithm deriving the wind speed from NRCS obtained at nadir incidence angle ($\theta = 0^\circ$) is used. A number of wind retrieval algorithms of various forms are known. For instance, the following NRCS model function at zero incidence angle $\sigma^\circ(U, 0^\circ)$ was proposed by (Chelton and McCabe 1985)

$$\sigma^\circ(U, 0^\circ)[dB] = 10(G_1 + G_2 \log_{10} U_{19.5}), \tag{2.6}$$

where G_1 and G_2 are the known parameters, $G_1 = 1.502$, $G_2 = -0.468$; $U_{19.5}$ is the wind speed at 19.5 m above the water surface. Comparison of the wind retrieval algorithms is presented in (Chelton et al. 2001; Gu et al. 2011; Schöne and Eickschen 2000; Zieger 2010).

Since coefficients in a geophysical model function are calculated by averaging NRCS values from a large-size data set, the NRCS model function does not consider local wind, wave, and temperature features, NRCS sampling variability, and instrumental measurement errors that do not allow to achieve accuracy of the wind speed and direction measurement better than mentioned above. Instrumental measurement noise is estimated at 0.2 dB, which corresponds to an uncertainty in wind speed of only 0.5 m/s (Stoffelen 1998). The relative precision of an airborne scatterometer calibration is generally around 0.2 dB, and absolute accuracy is about

1 dB (Nghiem et al. 1997). The accuracy of backscatter measurements also depends on the number of independent samples. Increasing this number, up to 5,000 for each cell as reported in (Nghiem et al. 1995), reduces the statistical fluctuation of detected power effectively. The presence of large waves affects backscatter most visibly at light winds because of the superposition of a large-scale roughness caused by a swell on wind-generated roughness. In that case, microwave NRCS increases at medium incidence angles and decreases at near-nadir angles by several decibels (5–6 dB), as reported in (Nghiem et al. 1995; Hwang et al. 1998). It may lead to overestimation of wind speed by several meters per second, and the measured wind direction may be considerably different from both average wind direction and principal wave directions. However, there is no significant impact of swell on backscatter measurements at moderate wind conditions (Hwang et al. 1998).

Thus, the scatterometer having an antenna with the inclined beam allows measuring the NRCS azimuth curve of a water surface and provides retrieval of both wind speed and wind direction over water. A scatterometer equipped with a nadir-looking antenna allows measuring nadir NRCS and estimating sea-surface wind speed but provides no information on the wind direction.

References

Bentamy A, Grodsky SA, Carton JA, Croizé-Fillon D, Chapron B (2012) Matching ASCAT and QuikSCAT winds. J Geophys Res 117(C02011): 1–15, https://doi.org/10.1029/2011JC007479

Carswell JR, Carson SC, McIntosh RE, Li FK, Neumann G, McLaughlin DJ, Wilkerson JC, Black PG, Nghiem SV (1994) Airborne scatterometers: investigating ocean backscatter under low- and high-wind conditions. Proc IEEE 82(12):1835–1860

Chelton DB, McCabe PJ (1985) A review of satellite altimeter measurement of sea surface wind speed: with a proposed new algorithm. J Geophys Res 90(C3):4707–4720

Chelton DB, Ries JC, Haines BJ, Fu L-L, Callahan PS (2001) Satellite altimetry. In: Fu L-L, Cazenave A (eds) Satellite altimetry and Earth sciences. A handbook of techniques and applications, vol 69. Int Geophys Series, Academic Press, San Diego, pp 1–131

Du Y, Vachon PW, Wolf J (2002) Wind direction estimation from SAR images of the ocean using wavelet analysis. Can J Remote Sens 28(3):498–509

Feindt F, Wismann V, Alpers W, Keller WC (1986) Airborne measurements of the ocean radar cross section at 5.3 GHz as a function of wind speed. Radio Sci 21(5):845–856

Fernandez DE, Carswell JR, Frasier S, Chang PS, Black PG, Marks FD (2006) Dual-polarized C- and Ku-band ocean backscatter response to hurricane-force winds. J Geophys Res 111 (C08013):1–17, https://doi.org/10.1029/2005JC003048

Giovanangeli J-P, Bliven LF, Calve OL (1991) A wind-wave tank study of the Azimuthal response of a Ka-Band scatterometer. IEEE Trans Geosci Remote Sens 29(1):143–148

Gu Y, Liu Y, Xu Q, Liu Y, Liu X, Ma Y (2011) A new wind retrieval algorithm for Jason-1 at high wind speeds. Int J Remote Sens 32(5):1397–1407

Hans P (1987) Auslegung und Analyse von satellitengetragenen Mikrowellensensorsystemen zur Windfeldmessung (Scatterometer) über dem Meer und Vergleich der Meßverfahren in Zeit- und Frequenzebene. Von der Fakultät 2 Bauingenieur- und Vermessungswesen der Universität Stuttgart zur Erlangung der Würde eines Doktor-Ingenieurs genehmigte Abhandlung. Institut für Navigation der Universität Stuttgart, Stuttgart, S 225

Hasager CB, Badger M, Peña A, Larsén XG, Bingöl F (2011) SAR-based wind resource statistics in the baltic sea. Remote Sens 3:117–144

Hildebrand PH (1994) Estimation of sea-surface wind using backscatter cross-section measurements from airborne research weather radar. IEEE Trans Geosci Remote Sens 32(1):110–117

Hwang PA, Teague WJ, Jacobs GA, Wang DW (1998) A statistical comparison of wind speed, wave height and wave period derived from satellite altimeters and ocean buoys in the Gulf of Mexico Region. J Geophys Res 103(C):10451–10468

Isoguchi O, Shimada M (2009) An L-Band ocean geophysical model function derived from PALSAR. IEEE Trans Geosci Remote Sens 47(7):1925–1936

Johannessen OM, Alexandrov VYu, Frolov IYe, Sandven S, Pettersson LH, Bobylov LP, Kloster K, Smirnov VG, Mironov YeU, Babich NG (2007) Remote sensing of sea ice in the Northern sea route: studies and applications. Springer-Praxis Publishing, Chichester, p 472

Karaev VYu, Kanevsky MV (1999) On the problem of the radar determination of the sea surface244 parameters. Exploration of the Earth from Space 4: 14–20, in Russian

Komen GJ, Cavaleri L, Donelan M, Hasselmann K, Hasselmann S, Janssen PAEM (1994) Dynamics and Modelling of Ocean Waves. Cambridge University Press, Cambridge, p 532

Li L, Heymsfield G, Carswell J, Schaubert D, McLinden M, Vega M, Perrine M (2011) Development of the NASA high-altitude imaging wind and rain airborne profiler. In: Proceedings of aerospace conference, Big Sky, MT, USA, 5–12 Mar 2011, pp 1–8

Long DG, Donelan MA, Freilich MH, Graber HC, Masuko H, Pierson WJ, Plant WJ, Weissman D, Wentz F (1996) Current progress in ku-band model functions. Brigham Young University, USA, Tech. Rep. MERS 96–002, p 88

Long MW (2001) Radar reflectivity of land and sea. Artech House, New York, p 534

Masuko H, Okamoto K, Shimada M, Niwa S (1986) Measurement of microwave backscattering signatures of the ocean surface using X band and Ka band airborne scatterometers. J Geophys Res 91(C11):13065–13083

Melnik YA (1980) Radar methods of the earth exploration. Sovetskoye Radio, Moscow, USSR, p 264, in Russian

Money DG, Mabogunje A, Webb D, Hooker M (1997) Sea clutter power spectral lineshape measurements. In: Proceedings of Radar'97, Edinburgh, UK, 14–16 Oct 1997, pp 85–89

Moore RK, Fung AK (1979) Radar determination of winds at sea. Proc IEEE 67(11):1504–1521

Moore RK, Jones WL (2004) Satellite scatterometer wind vector measurements: the legacy of the seasat satellite scatterometer. IEEE Geosci Remote Sens Newslett 132:18–32

Nghiem SV, Li FK, Lou SH, Neumann G, McIntosh RE, Carson SC, Carswell JR, Walsh EJ, Donelan MA, Drennan WM (1995) Observations of ocean radar backscatter at Ku and C bands in the presence of large waves during the Surface Wave Dynamics Experiment. IEEE Trans Geosci Remote Sens 33(3):708–721

Nghiem SV, Li FK, Neumann G (1997) The dependence of ocean backscatter at K_u-band on oceanic and atmospheric parameters. IEEE Trans Geosci Remote Sens 35:581–600

Nielsen SN, Long DG (2009) A wind and rain backscatter model derived from AMSR and SeaWinds data. IEEE Trans Geosci Remote Sens 47(6):1595–1606

Ouchi K (2000) A theory on the distribution function of backscatter radar cross section from ocean waves of individual wavelength. IEEE Trans Geosci Remote Sens 38(2):811–822

Plant WJ (2003) Microwave sea return at moderate to high incidence angles. Wave Random Media 13(4):339–354

Robbe N, Hengstermann T (2006) Remote sensing of marine oil spills from airborne platforms using multi-sensor systems. WIT Trans Ecol Envir 95:347–355

Schöne T, Eickschen S (2000) Wind speed and SWH calibration for radar altimetry in the North Sea. In: Proceedings of the ERS-Envisat symposium, Gothenburg, Sweden, 16–20 Oct 2000, p 8

Snoeijl P, Van Halsema D, Oost WA, Calkoen C, Jaehne B, Vogelzang J (1992) Microwave backscatter measurements made from the Dutch ocean research tower 'Noordwijk' compared with model predictions. In: Proceedings of IGARSS'92, Houston, TX, USA, 26–29 May 1992, pp 696–698

Spencer MW, Graf JE (1997) The NASA scatterometer (NSCAT) mission. Backscatter 8(4):18–24

Stoffelen A (1998) Scatterometry. Thesis. Universiteit Utrecht, Utrecht, p 209

Takeyama Y, Ohsawa T, Kozai K, Hasager CB, Badger M (2012) Effectiveness of weather research and forecasting wind direction for retrieving coastal sea surface wind from synthetic aperture radar. Wind Energ. https://doi.org/10.1002/we.1526

Ulaby FT, Moore RK, Fung AK (1982) Microwave Remote Sensing: Active and Pasive, vol 2: Radar remote sensing and surface scattering and emission theory. Addison-Wesley, London, p 1064

Wackerman CC, Rufenach CL, Shuchman RA, Johannessen JA, Davisdon KL (1996) Wind vector retrieval using ERS-1 synthetic aperture radar imagery. IEEE Trans Geosci Remote Sens 34 (6):1343–1352

Ward KD, Tough RJA, Watts S (2008) Sea clutter: Scattering, the K distribution and radar performance. Institution of Engineering and Technology, London, p 450

Wismann V (1989) Messung der Windgeschwindigkeit über dem Meer mit einem flugzeugge-tragenen 5.3 GHz Scatterometer. Dissertation zur Erlangung des Grades eines Doktors der Naturwissenschaften, Universität Bremen, Bremen, Germany, S 119

Zieger S (2010) Long term trends in ocean wind speed and wave height. Thesis. Swinburne University of Technology, Melbourne, p 177

Chapter 3
FM-CW Demonstrator System as an Instrument for Measuring Sea-Surface Backscattering Signature and Wind

3.1 FM-CW Demonstrator System

In the field of airborne earth observation, there is growing interest in small, cost-effective radar systems. Such radar systems should consume little power and be small enough to be mounted on light, possibly even unmanned, aircraft. FM-CW radar systems are generally compact, relatively cheap, and they consume little power. Consequently, FM-CW radar technology seems to be of interest to civil airborne earth observation, particularly in combination with high resolution SAR techniques.

A Dutch airborne Ka-band FM-CW demonstrator system based on off-the-shelf components has been designed at the International Research Center for Telecommunications-Transmission and Radar of Delft University of Technology to prove the feasibility of an FM-CW SAR under operational circumstances and to provide experimental test data (Meta et al. 2004).

This small, low-cost FM-CW demonstrator system has been fitted on a motor-glider standard pod that is attached under the wing. The pod has a diameter of about 35 cm and a length of 80 cm, excluding the aerodynamic fairings. The maximum payload of the pod is 50 kg. The demonstrator system has been fitted in such a pod as shown in Fig. 3.1. A Stemme S10 motor-glider, a relatively cheap test platform, has been used for experimental flights with the pods attached under the wings.

Over land, the demonstrator operates in a strip-map mode. The antenna axis is directed to the right side at a mounting incidence angle of 65°. The resolution is 1 m in range direction as well as in azimuth direction. Further specifications of the demonstrator system are summarized in Table 3.1.

In order to be able to estimate the position and attitude of the motor-glider, three gyroscopes, a tri-axial accelerometer and a Global Positioning System (GPS) receiver are attached to the demonstrator system. With aid of GPS readings, a position estimate can be updated from motion data. The GPS receiver outputs its position every second.

© The Author(s), under exclusive license to Springer Nature Switzerland AG 2021   13
A. Nekrasov, *Foundations for Innovative Application of Airborne Radars*,
SpringerBriefs in Earth Sciences, https://doi.org/10.1007/978-3-030-62942-7_3

Fig. 3.1 FM-CW demonstrator system installed in the standard pod

Table 3.1 Specifications of the FM-CW demonstrator system

Carrier frequency	35 GHz
Frequency sweep	500 MHz
Sweep repetition frequency	1,000/500 Hz
Modulation	Sawtooth/triangular
Transmitted power	18 dBm (0.063 W)
IF band	0–2.5 MHz
Antenna type	Horn-Lens
Antenna gain	24 dB
Antenna isolation	52 dB
Beam width azimuth/elevation	$6°/28°$
Platform velocity	30 m/s
Altitude	≤ 300 m
Maximum range	730 m

A digital camera has been added to the system in order to compare the radar images with pictures of the imaged area. During flights, the demonstrator system can be controlled and monitored from the cockpit with a pocket PC through a standard network connection. A more detailed description of the FM-CW demonstrator system is given in (Meta et al. 2004; Nekrasov et al. 2004).

3.2   Measuring Water-Surface Backscattering Signature by the FM-CW Demonstrator System

An aircraft equipped with the FM-CW demonstrator system operating in SAR mode performs the rectilinear track flight under measurements. Azimuth NRCS curve can be obtained using a circle track flight for a scatterometer with an inclined one-beam fixed-position antenna or a rectilinear track flight for a scatterometer with a rotating antenna (Carswell et al. 1994; Li et al. 2011; Masuko et al. 1986; Wismann 1989). Since the FM-CW demonstrator system has a fixed-position antenna, and since the system works in a scatterometer mode, the water-surface backscattering signature measurements should be completed during circle track flight.

As the radar system is installed on a motor-glider pod attached under the right wing, and the antenna beam axis is directed to the right side at an incidence (mounting) angle θ_m of 65° (at a straight flight), a clockwise circle flight should be completed with a right roll (Fig. 3.2).

Let a horizontal circle flight with the speed V and right roll γ_{fa} (right roll is positive) at some altitude H above mean sea surface be completed. Then, the current antenna incidence angle θ will depend on the roll angle of the flying apparatus γ_{fa} as follows

$$\theta = \theta_m - \gamma_{fa}, \tag{3.1}$$

and two distinct cases of the circle flight geometry for measuring the water-surface backscattering signature may occur. Figure 3.2a shows the case of $R_g < R_{t.fa}$, and Fig. 3.2b presents another case of $R_g \geq R_{t.fa}$, where R_g is the ground range (Nekrasov and Hoogeboom 2005)

$$R_g = \frac{H}{\tan \theta} = \frac{H}{\tan(\theta_{a.m} - \gamma_{fa})}, \tag{3.2}$$

and $R_{t.fa}$ is the radius of the aircraft turn (Mamayev et al. 2002)

$$R_{t.fa} = \frac{V^2}{g \tan|\gamma_{fa}|}, \tag{3.3}$$

where g is the acceleration of gravity, $g = 9.81$ m/s^2.

The radius of turn for a selected cell middle point $R_{t.c}$ for both cases of Fig. 3.2 is described by the following expression

$$R_{t.c} = |R_g - R_{t.fa}|. \tag{3.4}$$

Time of the aircraft 360° turn $T_{360°}$ is given by (Mamayev et al. 2002)

$$T_{360°} = \frac{2\pi V}{g \tan|\gamma_{fa}|}. \tag{3.5}$$

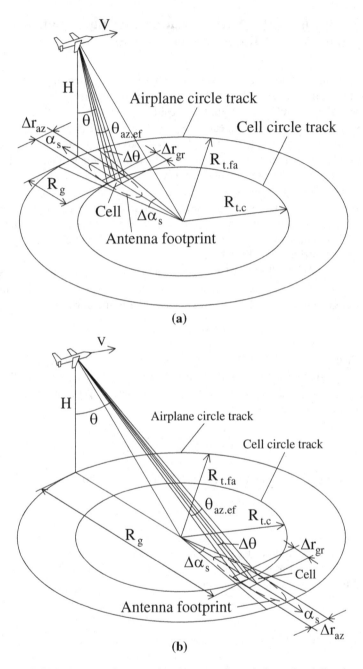

Fig. 3.2 Circle flight geometry for measuring the sea-surface wind speed and direction: **a** in the case of $R_g < R_{t,fa}$; **b** in the case of $R_g \geq R_{t,fa}$

Usually, 360° azimuth space is divided into 72 or 36 sectors under the circle NRCS measurement. Azimuth size of a sector observed is 5° or 10°, respectively. The middle azimuth of the sector is the azimuth of the sector observed. The azimuth size of a sector relative to the center point of circle of the airplane track is $\Delta\alpha_s$, and the middle azimuth of a sector is α_s. The NRCS samples obtained from a sector and averaged over all measurement values in that sector give the NRCS value $\sigma°(\theta, \psi_s)$ corresponding to the real observation azimuth angle of the sector ψ_s that is (Nekrasov 2011)

$$\psi_s = \psi_{\psi_s} + \psi_b \pm 360°, \tag{3.6}$$

where ψ_{ψ_s} is the aircraft course corresponding to the real observation azimuth angle of the sector, where ψ_b is the azimuth direction of the antenna beam relative to the aircraft course corresponding to the real observation azimuth angle of the sector (right position is positive), $\pm 360°$ means that if the summation result is 360° or greater, it should be corrected by minus 360°, but if it is negative, it should be corrected by plus 360° because of the summation result always should belong to the interval $[0°, 360°[$.

As the aircraft performs a clockwise turn, the real observation azimuth angles of the sector beginning $\psi_{s.b}$ and of the sector end $\psi_{s.e}$ are

$$\psi_{s.b} = \psi_s - \Delta\alpha_s/2 \pm 360°, \tag{3.7}$$

$$\psi_{s.e} = \psi_s + \Delta\alpha_s/2 \pm 360°. \tag{3.8}$$

Time of a sector view T_s and number of samples N_s that can be obtained from the observed sector are presented by the following expressions (Nekrasov 2010a)

$$T_s = T_{360°} \frac{\Delta\alpha_s}{360°}, \tag{3.9}$$

$$N_s = \frac{T_s V}{0.5a}, \tag{3.10}$$

where a is the antenna length in the direction of flight.

The size of a selected cell is defined by azimuth Δr_{az} and ground-range Δr_{gr} resolutions, which are

$$\Delta r_{az} = \frac{H\theta_{az.ef}}{\cos\theta}, \tag{3.11}$$

$$\Delta r_{gr} = \frac{H\Delta\theta}{\cos^2\theta}, \tag{3.12}$$

where $\theta_{az.ef}$ is the effective (two-way) azimuth beamwidth, $\Delta\theta$ is the angular (two-way) resolution in the vertical plane. By selecting proper range resolution, $\Delta\theta$ is set to 4°, to comply with the required power budget. It is assumed that the NRCS values are equal at any point of the selected cell. Using $sin(x)/x$ approximation for the major lobe of the antenna beam of the FM-CW demonstrator system, the effective (two-way) azimuth beamwidth of 4.4°, and effective (two-way) antenna elevation beamwidth $\theta_{el.ef}$ of 20.6° have been calculated (antenna length a is 0.137 m, antenna width b is 0.018 m).

The power P_r received from the resolution cell is given by

$$P_r = \frac{P_t G^2 \lambda^2 \sigma_{cell} \cos^4 \theta}{(4\pi)^3 H^4 L_s} \int\limits_{-0.5\theta_{az.ef}}^{0.5\theta_{az.ef}} \int\limits_{-0.5\Delta\theta}^{0.5\Delta\theta} g^2(\theta', \phi') d\theta' d\phi', \tag{3.13}$$

where P_t is the transmitted power, G is the antenna gain, λ is the carrier wavelength, σ_{cell} is the radar cross section of the selected cell, L_s is the system losses, $g(\theta', \phi')$ is the antenna directivity pattern, θ' and ϕ' are the incidence angle and the azimuth angle relative to the antenna beam axis, respectively.

The radar cross section of the selected cell is as follows

$$\sigma_{cell} = \Delta r_{gr} \Delta r_{az} \sigma^\circ, \tag{3.14}$$

where σ° is the NRCS of the target (reflecting surface).

Antenna pattern can be separated into two components (Ulaby et al. 1982)

$$g(\theta', \phi') = g(\theta') g(\phi'), \tag{3.15}$$

where $g(\theta')$ and $g(\phi')$ are the pattern factors in the across-track and along-track directions, respectively. Using (3.15), a $sin(x)/x$ approximation for the major lobe of the antenna beam, and circle flight geometry of measurement (Fig. 3.2) with the FM-CW demonstrator system under chosen angular resolutions in the vertical and horizontal planes, the following solution has been obtained

$$\int\limits_{-0.5\theta_{az.ef}}^{0.5\theta_{az.ef}} \int\limits_{-0.5\Delta\theta}^{0.5\Delta\theta} g^2(\theta', \phi') d\theta' d\phi'$$

$$= \int\limits_{-0.5\theta_{az.ef}}^{0.5\theta_{az.ef}} g^2(\phi') d\phi' \int\limits_{-0.5\Delta\theta}^{0.5\Delta\theta} g^2(\theta') d\theta' = 0.81$$

Thus, the power received from one resolution cell can be written in the following form

$$P_r = \frac{0.81 P_t G^2 \lambda^2 \Delta r_{gr} \Delta r_{az} \sigma^\circ \cos^4 \theta}{(4\pi)^3 H^4 L_s}, \tag{3.16}$$

where 0.81 is the beam shape correction factor for the chosen angular resolutions (4° resolution in the vertical plane and 4.4° resolution in the horizontal plane), and then, using (3.16), averaged NRCS value $\sigma^\circ(\theta, \psi_s)$ obtained by the FM-CW demonstrator system from an appropriate sector can be presented as follows (Nekrasov and Hoogeboom 2005)

$$\sigma^\circ(\theta, \psi_s) = \frac{0.81 P_t G^2 \lambda^2 \Delta r_{gr} \Delta r_{az} \bar{P}_r(\theta, \psi_s) \cos^4 \theta}{(4\pi)^3 H^4 L_s},$$ (3.17)

where $\bar{P}_r(\theta, \psi_s)$ is the average power received from the sector.

Since the measured NRCS data set is in fact discrete, each NRCS value obtained is $\sigma^\circ(\theta, \psi_{s.i})$, where $i = \overrightarrow{1, N}$, N is the number of sectors observed during the airplane turn for 360° at the NRCS measurement, $N = 360°/\Delta\alpha_s$.

Thus, to obtain an azimuth NRCS curve of the water surface under aircraft circle flight by the FM-CW demonstrator system operating in scatterometer mode, the measurement should be performed in accordance with a scheme of Fig. 3.3.

The flight should be performed so as the circle track is located near a center point of the studied area of the water surface. Recommended main flight parameters to perform measuring the water-surface backscattering signature with the FM-CW demonstrator system mounted on a Stemme S10 motor-glider are:

altitude $H = 300$ m;
speed of flight $V = 25$ m/s;
right roll $\gamma_{fa} = 20°$;
radius of the airplane turn $R_{t.fa} = 175$ m;
time of the airplane turn for 360° $T_{360°} = 44$ s.

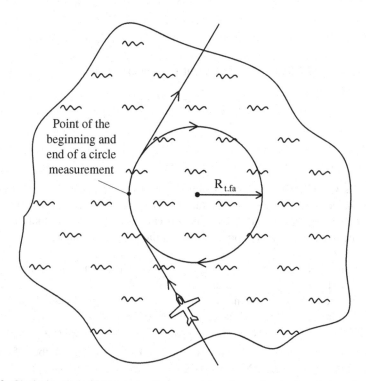

Fig. 3.3 Clockwise circle flight scheme for measuring water surface backscattering signature

The measurement is started when a stable horizontal circle flight at the given altitude, speed of flight, roll and pitch has been established. The measurement is finished when the azimuth of the measurement start is reached. To obtain a greater number of NRCS samples for each sector observed, several consecutive full circle 360° turns may be done.

3.3   Wind Retrieval from Azimuth NRCS Data Obtained with the FM-CW Demonstrator System

As the azimuth NRCS data is obtained from the azimuth sector whose width may exceed several degrees, the influence of azimuth width on the NRCS value should also be considered.

Let the sea-surface backscattering for medium incidence angles be of the form (2.1). As the azimuth NRCS curve (2.1) is smooth, and taking into account the azimuth angular size $\Delta\alpha$ of a sector, the expression for its NRCS $\sigma°(U, \theta, \alpha, \Delta\alpha)$ can be presented in the following form (Nekrassov 1999)

$$
\begin{aligned}
\sigma°(U, \theta, \alpha, \Delta\alpha) &= \frac{1}{\Delta\alpha} \int_{\alpha-0.5\Delta\alpha}^{\alpha+0.5\Delta\alpha} \sigma°(U, \theta, \alpha')d\alpha' \\
&= A(U, \theta) + k_1(\Delta\alpha)B(U, \theta)\cos\alpha \\
&\quad + k_2(\Delta\alpha)C(U, \theta)\cos(2\alpha),
\end{aligned}
\tag{3.18}
$$

where $k_1(\Delta\alpha)$ and $k_2(\Delta\alpha)$ are the coefficients dependent on the azimuth sector width

$$
k_1(\Delta\alpha) = \frac{2\sin(0.5\Delta\alpha)}{\Delta\alpha},
\tag{3.19}
$$

$$
k_2(\Delta\alpha) = \frac{\sin\Delta\alpha}{\Delta\alpha}.
\tag{3.20}
$$

Figure 3.4 and Table 3.2 show the dependence of coefficients $k_1(\Delta\alpha)$ and $k_2(\Delta\alpha)$ on azimuth sector width (azimuth resolution). They demonstrate that in case of the selected cell being narrow enough in the vertical plane the NRCS model function (2.1) can be used without any correction when the azimuth angular size of a cell is up to $15° - 20°$.

As measured NRCS data set is also a function of the wind speed, each NRCS value obtained $\sigma°(\theta, \psi_{s.i})$ is considered now as $\sigma°(U, \theta, \psi_{s.i})$.

Let angle between the up-wind direction and the first NRCS azimuth $\psi_{s.1}$ be α, the sector width be $\Delta\alpha_s$, and so the measured NRCSs $\sigma°(U, \theta, \psi_{s.1})$, $\sigma°(U, \theta, \psi_{s.2})$, ..., $\sigma°(U, \theta, \psi_{s.N})$ be the same as $\sigma°(U, \theta, \alpha)$, $\sigma°(U, \theta, \alpha + \Delta\alpha)$,..., $\sigma°(U, \theta, \alpha + (N-1)\Delta\alpha)$, respectively, where $i = \overline{1, N}$, N is the number of sectors composing the 360° azimuth NRCS curve, $N = 360°/\Delta\alpha_s$. Then, in a general case,

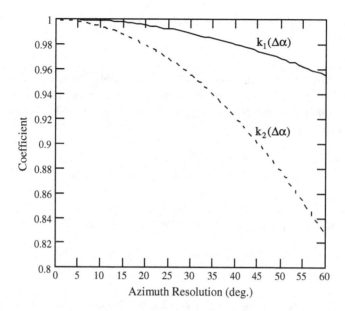

Fig. 3.4 Dependence of coefficients $k_1(\Delta\alpha)$ and $k_2(\Delta\alpha)$ on azimuth angular size of a sector

Table 3.2 Dependence of coefficients $k_1(\Delta\alpha)$ and $k_2(\Delta\alpha)$ on the azimuth angular size of a sector

$\theta_{a.h}$	5°	10°	15°	20°	30°	45°	60°
$k_1(\theta_{a.h})$	1	1	1	1	0.99	0.98	0.96
$k_2(\theta_{a.h})$	1	1	0.99	0.98	0.96	0.9	0.83

to find the wind speed and up-wind direction from the azimuth NRCS data set obtained clockwise the following system of N equations should be solved approximately using searching procedure within the ranges of discrete values of possible solutions

$$
\begin{cases}
\sigma^\circ(U,\theta,\alpha) = A(U,\theta) + B(U,\theta)\cos\alpha \\
\qquad + C(U,\theta)\cos(2\alpha), \\
\sigma^\circ(U,\theta,\alpha+\Delta\alpha_s) = A(U,\theta) \\
\qquad + B(U,\theta)\cos(\alpha+\Delta\alpha_s) \\
\qquad + C(U,\theta)\cos(2(\alpha+\Delta\alpha_s)), \\
\cdots\cdots\cdots\cdots\cdots\cdots\cdots\cdots\cdots\cdots\cdots\cdots\cdots\cdots \\
\sigma^\circ(U,\theta,\alpha+(N-2)\Delta\alpha_s) = A(U,\theta) \\
\qquad + B(U,\theta)\cos(\alpha+(N-2)\Delta\alpha_s) \\
\qquad + C(U,\theta)\cos(2(\alpha+(N-2)\Delta\alpha_s)), \\
\sigma^\circ(U,\theta,\alpha+(N-1)\Delta\alpha_s) = A(U,\theta) \\
\qquad + B(U,\theta)\cos(\alpha+(N-1)\Delta\alpha_s) \\
\qquad + C(U,\theta)\cos(2(\alpha+(N-1)\Delta\alpha_s)),
\end{cases}
\tag{3.21}
$$

and navigation wind direction can be found as following

$$\psi_w = \psi_{s.1} - \alpha \pm 180°. \tag{3.22}$$

As an entire 360° azimuth NRCS data set is available, the wind vector recover procedure (3.21) can be simplified and thus sped up significantly because of the 360° azimuthally average NRCS $\sigma°_{360°}(U, \theta)$ based on the NRCS model function (2.1) can be presented in the following form (Nekrassov 2002)

$$\sigma°_{360°}(U, \theta) = \frac{1}{2\pi} \int_0^{2\pi} \sigma°(U, \theta, \alpha)d\alpha = A(U, \theta), \tag{3.23}$$

and then, the wind speed equation can be written down as following

$$U = \left(\frac{A(U, \theta)}{a_0(\theta)}\right)^{1/\gamma_0(\theta)} = \left(\frac{\sigma°_{360°}(U, \theta)}{a_0(\theta)}\right)^{1/\gamma_0(\theta)}. \tag{3.24}$$

Then, taking into account that the azimuth NRCS data set is discrete, the following algorithm to retrieve the wind vector over the water surface from the measured NRCS data set can be proposed.

The wind speed can be found from the following equation (Nekrasov and Hoogeboom 2005)

$$U = \left(\frac{\sum\limits_{i=1}^{360°/\Delta\alpha_s} \sigma°(U, \theta, \psi_{s.i})}{\frac{360°}{\Delta\alpha_s} a_0(\theta)}\right)^{1/\gamma_0(\theta)}. \tag{3.25}$$

Then, the up-wind direction can be estimated using a searching procedure to find the main maximum value from the azimuth NRCS set

$$\sigma°_{max} = \max_{i=1, \, 360°/\Delta\alpha_s} \{\sigma°(U, \theta, \psi_{s.i})\}, \tag{3.26}$$

and the azimuth of the NRCS $\psi_{\sigma°_{max}}$ corresponding to it should be found. After that, the navigation wind direction can be found from (2.5).

Unfortunately, the significant statistical variation of the NRCS data may take place, even after the NRCS integration in azimuth, and it may lead to a considerable error in the wind direction estimation. Therefore, to reduce the influence of the statistical variation of the NRCS data on the wind vector retrieval, the least squares method (Bronshtein et al. 2004) can be used to obtain a more precise location of the NRCS main maximum value, when a larger NRCS set is used. The NRCS model function (2.1) at the measured wind speed obtained from (3.25) is regarded to be the best approximation of sampled NRCS values within $\pm 360°$ relative to its principal maximum in azimuth, and then, the least value of summation results by the least squares method S_{min} can be written as follows (Nekrasov and Hoogeboom 2005)

$$S_{min} = \min_{\substack{j=0, \frac{360°}{\Delta\alpha_s}}} \left\{ \sum_{i=j-\frac{60°}{\Delta\alpha_s}}^{j+\frac{60°}{\Delta\alpha_s}} \left(\sigma°(U, \theta, \psi_{s,j}) - \sigma°(U, \theta, \psi_i) \right)^2 \right\}. \tag{3.27}$$

The azimuth of the NRCS $\psi_{s,j}$ corresponding to the least value of summation results S_{min} will be $\psi_{\sigma°_{max}}$, and wind direction can be found from (2.5).

Alternatively, knowing the wind speed from (3.25), the up-wind direction also can be estimated using (3.21) and (3.22). It provides minimum error in the wind direction retrieval.

3.4   Simulation of Wind Vector Retrieval from Azimuth NRCS Data

To investigate the operational capability of the proposed measuring wind algorithm, a simulation of the wind vector retrieval based on an appropriate NRCS model function needs to be performed.

The model function can be calculated from a measured NRCS data set obtained under collocated NRCS and wind measurement at appropriate frequency and polarization, and for various wind speed, incidence, and azimuth angles. If such a sufficient data set is available, the Fourier coefficients $A(U, \theta)$, $B(U, \theta)$, and $C(U, \theta)$ in the NRCS model function (2.1) can be estimated as a function of both the incident angle and the wind speed as follows (Ulaby et al. 1982)

$$A(U, \theta) = \frac{\sigma°_u(U, \theta) + \sigma°_d(U, \theta) + 2\sigma°_c(U, \theta)}{4}, \tag{3.28}$$

$$B(U, \theta) = \frac{\sigma°_u(U, \theta) - \sigma°_d(U, \theta)}{2}, \tag{3.29}$$

$$C(U, \theta) = \frac{\sigma°_u(U, \theta) + \sigma°_d(U, \theta) - 2\sigma°_c(U, \theta)}{4}, \tag{3.30}$$

where $\sigma°_u(U, \theta)$, $\sigma°_d(U, \theta)$, and $\sigma°_c(U, \theta)$ are the up-wind, down-wind, and cross-wind NRCS estimates, respectively. It means that at least the NRCS data set for the up-wind, down-wind, and cross-wind azimuth directions should be provided to recover the coefficients in the geophysical model function (2.1) but that data set should be large enough to provide good statistical sampling.

As the azimuth NRCS curve (1.1) is smooth, the NRCS values from sectors nearer to considered up-wind, down-wind, and cross-wind sectors also can be used to calculate the Fourier coefficients $A(U, \theta)$, $B(U, \theta)$, and $C(U, \theta)$ so as to increase the number of NRCS samples. That is apparent from (3.18)–(3.20) and Table 3.2.

Since such NRCS data set as well as the coefficients for the Ka-band NRCS model function of the form (2.1) were not available but necessary for the wind vector retrieval simulation, another NRCS model function from (Masuko et al. 1986) has been used to develop the NRCS model function of the form (2.1) for incidence angles of $30° - 50°$. That model function is presented here in the modified (more convenient) form adapted to a 10-meter height of measurement currently used

$$\sigma° = 10^Q \left(\frac{U}{0.93} \right)^h, \tag{3.31}$$

where Q and h are the coefficients. Unfortunately, those coefficients were given only for up-, down-, and cross-wind directions (Q_u, h_u, Q_d, h_d, and Q_c, h_c) and for incidence angles of $30°$, $40°$, $50°$, and $60°$. Table 3.3 shows the values of the coefficients of the 34.43 GHz NRCS model function (2.1) for vertical transmit and receive polarization (Masuko et al. 1986).

At first, using up-, down-, and cross-wind curves of (3.31) for the incidence angles of $30°$, $40°$, and $50°$, the coefficients $a_0(\theta)$, $a_1(\theta)$, $a_2(\theta)$, $\gamma_0(\theta)$, $\gamma_1(\theta)$, and $\gamma_2(\theta)$ were obtained for those fixed incidence angles. Then, using approximate solutions to the systems of nonlinear equations, each composed of three equations for the same coefficient at the incidence angles of $30°$, $40°$, and $50°$, the following coefficients for the Ka-band NRCS model of form (2.1) for vertical transmit and receive polarization, incidence angles of $30°$ to $50°$ and wind speeds of 3–20 m/s were computed (Nekrasov and Hoogeboom 2005)

$$\begin{aligned}
a_0(\theta) &= 0.006036 - 0.0002031\theta + 0.0000001680\theta^2, \\
a_1(\theta) &= -0.007776 + 0.0004421\theta - 0.000005692\theta^2, \\
a_2(\theta) &= 0.001129 + 0.00001335\theta - 0.0000006895\theta^2, \\
\gamma_0(\theta) &= 4.902 - 0.198\theta + 0.002810\theta^2, \\
\gamma_1(\theta) &= 13.618 - 0.631\theta + 0.0076650\theta^2, \\
\gamma_2(\theta) &= 5.896 - 0.258\theta + 0.003550\theta^2,
\end{aligned} \tag{3.32}$$

Table 3.3 Coefficients of the 34.43 GHz NRCS model function of the form (2.1) for vertical transmit and receive polarization

θ	30°	40°	50°	60°
Q_u	−2.633	−2.800	−3.629	−3.962
h_u	1.488	1.285	1.791	1.884
Q_d	−2.307	−3.404	−4.120	−4.157
h_d	1.002	1.721	2.112	1.809
Q_c	−3.315	−4.333	−5.170	−4.897
h_c	1.668	2.155	2.600	2.125

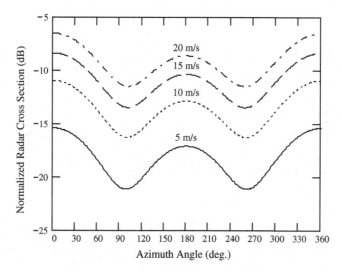

Fig. 3.5 Ka-band NRCS versus azimuth angle at wind speeds of 5, 10, 15, and 20 m/s at 10-meter height for incidence angle of 30° at vertical transmit and receive polarization

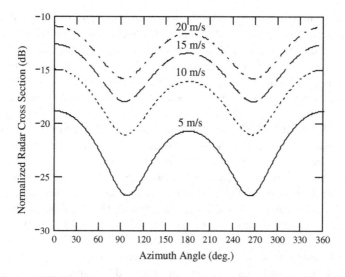

Fig. 3.6 Ka-band NRCS versus azimuth angle at wind speeds of 5, 10, 15, and 20 m/s at 10-meter height for incidence angle of 40° at vertical transmit and receive polarization

where θ is the incidence angle in degrees.

Figures 3.5, 3.6, 3.7 give the examples of NRCS curves of form (2.1) with coefficients (3.32) versus azimuth angle at wind speeds of 5, 10, 15, and 20 m/s at 10-meter for incidence angles of 30°, 40°, and 50°, respectively, at vertical transmit and receive polarization.

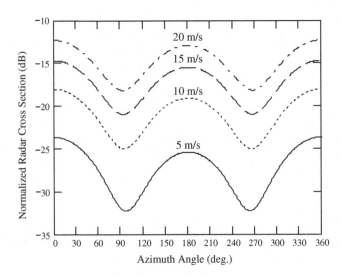

Fig. 3.7 Ka-band NRCS versus azimuth angle at wind speeds of 5, 10, 15, and 20 m/s at 10-meter height for incidence angle of 50° at vertical transmit and receive polarization

It is important to note that under the model development the original up-wind NRCS curve from (Masuko et al. 1986) at the incidence angle of 30° has been corrected. The down-wind NRCS values tended to exceed the up-wind NRCS value at lower wind speeds, as well as the down-wind NRCS value tended to exceed the cross-wind NRCS value at higher wind speeds. The correction is based on the wind-wave tank study of Ka-band NRCS at the incidence angle of 30° reported in (Giovanangeli et al. 1991).

Thus, the Ka-band NRCS model function required for the FM-CW demonstrator system is available now, and the operational capability of the measuring wind algorithm proposed can be proved by means of simulation.

The "measured" azimuth NRCS values were generated using Rayleigh Power (Exponential) distribution and (2.1) with the coefficients (3.32). Then, the proposed algorithm for recovering wind speed and direction (3.25)–(3.27) has been tested at wind speeds of 3–20 m/s under averaging of a various number of NRCS samples.

Figures 3.8, 3.9, 3.10, 3.11, 3.12, 3.13, 3.14, 3.15, 3.16, 3.17, 3.18 present three examples of the results obtained at the following conditions: wind speeds are 5, 10, and 15 m/s, true up-wind direction is 0°, altitude is 300 m, speed of flight is 25 m/s, right roll is 20°. At those conditions, the incidence angle is 45°, and time of the aircraft 360° turn is 44 s.

Figures 3.8, 3.12, and 3.16 show "measured" NRCS after averaging of 44 samples in a 1° azimuth sector (dotted traces) at the wind speeds of 5, 10, and 15 m/ s, respectively. Figures 3.9, 3.13, and 3.17 demonstrate "measured" NRCS after averaging of 220 samples in a 5° azimuth sector (dotted traces) at the same wind speeds, respectively. Figures 3.10, 3.14, and 3.18 present "measured" NRCS after

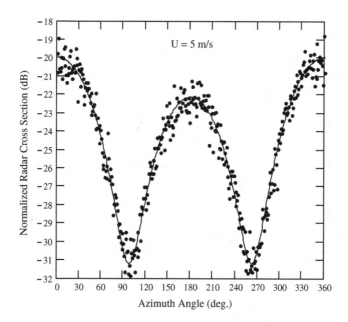

Fig. 3.8 "Measured" NRCS after averaging of 44 samples in a 1° azimuth sector (*dot trace*) and azimuth NRCS curve by model (2.1) with the coefficients (3.32) (*solid trace*) at wind speed of 5 m/s

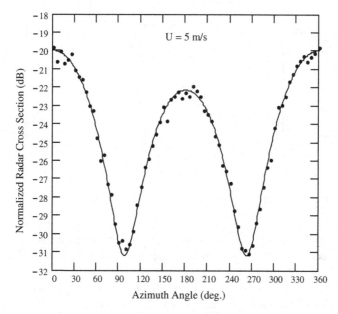

Fig. 3.9 "Measured" NRCS after averaging of 220 samples in a 5° azimuth sector (*dot trace*) and azimuth NRCS curve by model (2.1) with the coefficients (3.32) (*solid trace*) at wind speed of 5 m/s

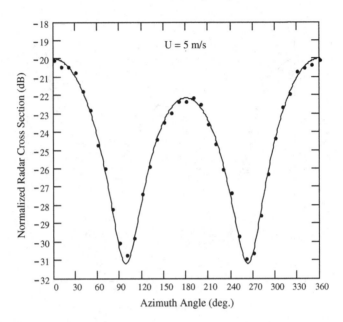

Fig. 3.10 "Measured" NRCS after averaging of 440 samples in a 10° azimuth sector (*dot trace*) and azimuth NRCS curve by model (2.1) with the coefficients (3.32) (*solid trace*) at wind speed of 5 m/s

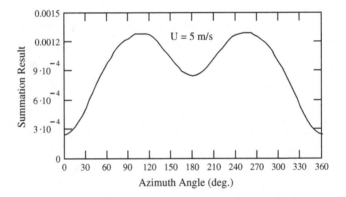

Fig. 3.11 Cumulative result by the least square method at wind speed of 5 m/s

averaging of 440 samples in a 10° azimuth sector (dot traces) also at the same wind speeds, respectively. Solid traces in those figures show the azimuth NRCS curves by model (2.1) with the coefficients (3.32).

Figures 3.11, 3.15, and 3.19 give the cumulative results by the least square method (3.27) at the same conditions at the wind speeds of 5, 10, and 15 m/s, respectively.

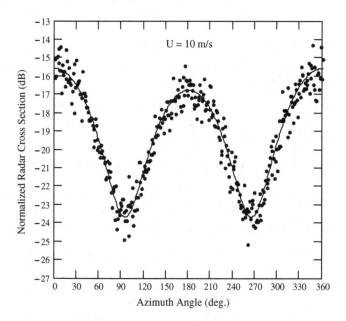

Fig. 3.12 "Measured" NRCS after averaging of 44 samples in a 1° azimuth sector (*dot trace*) and azimuth NRCS curve by model (2.1) with the coefficients (3.32) (*solid trace*) at wind speed of 10 m/s

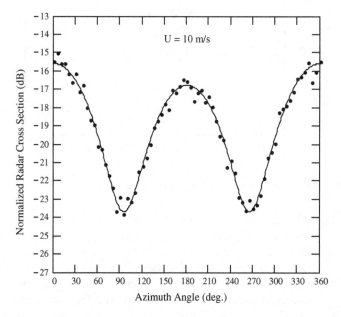

Fig. 3.13 "Measured" NRCS after averaging of 220 samples in a 5° azimuth sector (*dot trace*) and azimuth NRCS curve by model (2.1) with the coefficients (3.32) (*solid trace*) at wind speed of 10 m/s

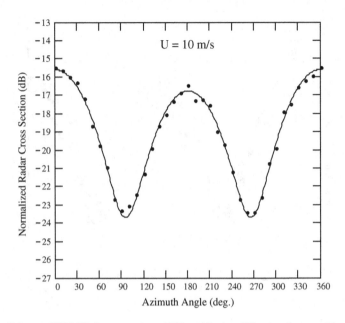

Fig. 3.14 "Measured" NRCS after averaging of 440 samples in a 10° azimuth sector (*dot trace*) and azimuth NRCS curve by model (2.1) with the coefficients (3.32) (*solid trace*) at wind speed of 10 m/s

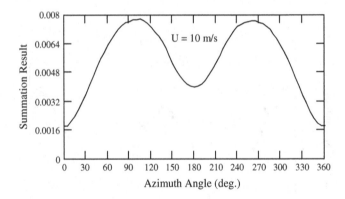

Fig. 3.15 Cumulative result by the least square method at wind speed of 10 m/s

Using (3.25), the "measured" wind speed is 4.98 m/s for the "true" wind speed of 5 m/s, and the "measured" up-wind directions are 0°, 0° and 0° for the 5° sector of averaging, the 10° sector of averaging, and the least squares method (3.27), respectively. For the "true" wind speed of 10 m/s, the "measured" wind speed is 9.95 m/s, and the "measured" up-wind directions are 5°, 0° and 0°. For the "true" wind speed of 15 m/s, the "measured" wind speed is 14.98 m/s, and the "measured" up-wind directions are 355°, 0° and 0°. The "measured" wind speeds are a little lower than the "true" wind speeds, no lower than 0.05 m/s.

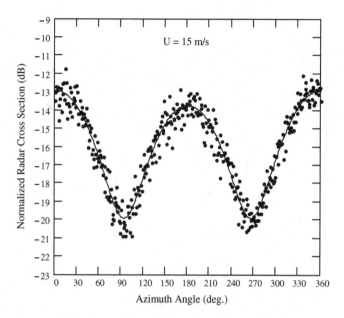

Fig. 3.16 "Measured" NRCS after averaging of 44 samples in a 1° azimuth sector (*dot trace*) and azimuth NRCS curve by model (2.1) with the coefficients (3.32) (*solid trace*) at wind speed of 15 m/s

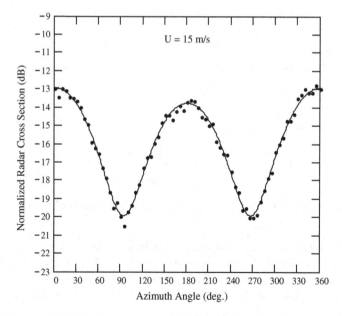

Fig. 3.17 "Measured" NRCS after averaging of 220 samples in a 5° azimuth sector (*dot trace*) and azimuth NRCS curve by model (2.1) with the coefficients (3.32) (*solid trace*) at wind speed of 15 m/s

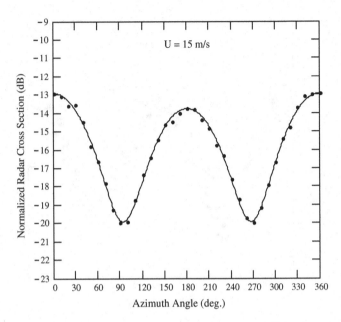

Fig. 3.18 "Measured" NRCS after averaging of 440 samples in a 10° azimuth sector (*dot trace*) and azimuth NRCS curve by model (2.1) with the coefficients (3.32) (*solid trace*) at wind speed of 15 m/s

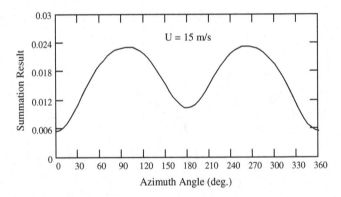

Fig. 3.19 Cumulative result by the least square method at wind speed of 15 m/s

Also, the influence of the systematic error in the NRCS on the wind measurement results has been studied. The positive error in the NRCS leads to overstating the wind speed value, but the negative error in the NRCS leads to understating the wind speed value. For instance, positive error in the NRCS of 0.5 dB gives the "measured" wind speed values of 5.33, 10.66, and 16.04 m/s, respectively, to the "true" wind speeds of 5, 10, and 15 m/s.

Those examples show us clearly the suitability of the algorithm proposed. The wind deviation (± 1.5 m/s) is well within the typical accuracy of the scatterometer wind measurement in general. The "measured" wind directions ($\pm 15°$) are also within the typical accuracy of a scatterometer measurement. Other simulation results, not presented here, have shown that the wind direction obtained by the least squares method is more accurate than the wind direction retrieved using (3.26) along with the NRCS averaging in the 5° or 10° azimuth sector.

As the power reflected from a sea surface varies, to reduce the statistical fluctuation of detected power a number of independent samples for each azimuth direction (cell) are integrated. The required number of integrated samples from a cell to discriminate up-wind and down-wind NRCS values with the ratio of 0.5 dB is approximately 315. In the worst-case scenario when the NRCS values are not from up-wind to down-wind directions, not equal but differ by 0.2 dB or 0.1 dB, the required number of integrated samples becomes 785 or 1,565, respectively (Nekrasov 2008a). Thus, use of the 360° azimuth NRSC curve will provide more precise wind vector estimation over water in comparison with the cases when the NRCS data only from some (three or four) azimuth directions are considered. Later, the coefficients (3.32) for the Ka-band NRCS model of form (2.1) has been justified in (Nekrasov et al. 2017).

3.5 Conclusions to the FM-CW Demonstrator System as an Instrument for Measuring Sea-Surface Backscattering Signature and Wind

The complete evaluation of the FM-CW demonstrator system has shown that the system is feasible for remotely measuring sea-surface backscattering signature and recovering the near-surface wind vector over water from the obtained NRCS azimuth curves.

It is desirable to provide the incidence angle of the selected cells $\theta \rightarrow 45°$ at measurement which is explained by better usage of the anisotropic properties of the water-surface scattering at medium incidence angles (Ulaby et al. 1982) as well as by power reasons. For water surface, NRCS falls radically as incidence angle increases and assumes different values for different conditions of sea state or water roughness while, for most other types of terrain, the NRCS decreases slowly with increase of the beam incidence angle (Kayton and Fried 1997). Otherwise, the incidence angle of the selected cells at least should be in the range of validity of the NRCS model function (2.1) and should be out of the "shadow" region of the water backscatter.

To perform the measurement, the FM-CW demonstrator system should operate in the scatterometer mode using the proposed algorithm. The measurement is started when a stable horizontal circle flight at the given altitude, speed of flight, roll and pitch has been established. The measurement is finished when the azimuth of the measurement start is reached. To obtain a greater number of NRCS samples for each sector observed, several consecutive full circle 360° turns may be done.

References

Carswell JR, Carson SC, McIntosh RE, Li FK, Neumann G, McLaughlin DJ, Wilkerson JC, Black PG, Nghiem SV (1994) Airborne scatterometers: investigating ocean backscatter under low- and high-wind conditions. Proc IEEE 82(12):1835–1860

Giovanangeli J-P, Bliven LF, Calve OL (1991) A Wind-Wave Tank Study of the Azimuthal Response of a Ka-Band Scatterometer. IEEE Trans Geosci Remote Sens 29(1):143–148

Kayton M, Fried WR (1997) Avionics Navigation Systems. Wieley, New York, p 773

Li L, Heymsfield G, Carswell J, Schaubert D, McLinden M, Vega M, Perrine M (2011) Development of the NASA high-altitude imaging wind and rain airborne profiler. In: Proceedings of Aerospace Conference, Big Sky, MT, 5–12 Mar 2011, pp 1–8

Mamayev VY, Sinyakov AN, Petrov KK, Gorbunov DA (2002) Air navigation and elements of navigation calculations. GUAP, Saint Petersburg, Russia, p 256, in Russian

Masuko H, Okamoto K, Shimada M, Niwa S (1986) Measurement of microwave backscattering signatures of the ocean surface using X band and Ka band airborne scatterometers. J Geophys Res 91(C11):13065–13083

Meta A, De Wit JJM, Hoogeboom P (2004) Development of a high resolution airborne millimeter wave FM-CW SAR. In: Proceedings of EuRAD. Amsterdam, Netherlands, 11–15 Oct 2004, pp 209–212

Nekrasov A (2008a) Measurement of the wind vector over sea by an airborne radar altimeter having an antenna with the different beamwidth in the vertical and horizontal planes. IEEE Geosci Remote Sens Lett 5(1):31–33

Nekrasov A (2010a) Airborne Doppler navigation system application for measurement of the water surface backscattering signature. In: Wagner W, Székely B (eds): ISPRS TC VII symposium: 100 Years ISPRS, Vienna, Austria, 2–4 Jul 2010, International archives of the photogrammetry, remote sensing and spatial information Sciences, 2010, vol. XXXVIII, part 7A, pp 163–168

Nekrasov A (2011) Airborne weather radar application for measurement of the water surface backscattering signature. In: Proceedings of RADAR 2011, Chengdu, China, 24–27 Oct 2011, p 4

Nekrasov A, De Wit JJM, Hoogeboom P (2004) FM-CW millimeter wave demonstrator system as a sensor of the sea surface wind vector. IEICE Electron Express 1(6):137–143

Nekrasov A, Hoogeboom P (2005) A Ka-band backscatter model function and an algorithm for measurement of the wind vector over the sea surface. IEEE Geosci Remote Sens Lett 2(1):23–27

Nekrassov A (1999) Sea surface wind vector measurement by airborne scatterometer having wide-beam antenna in horizontal plane. In: Proceedings of IGARSS'99, Hamburg, Germany, 28 June-2 Jul 1999, vol 2, pp 1001–1003

Nekrassov A (2002) On airborne measurement of the sea surface wind vector by a scatterometer (altimeter) with a nadir-looking wide-beam antenna. IEEE Trans Geosci Remote Sens 40 (10):2111–2116

Nekrasov A, Popov D, Schünemann K (2017) A Ka-band geophysical model function. Microw Rev 23(2):20–23

Ulaby FT, Moore RK, Fung AK (1982) Microwave Remote Sensing: Active and Pasive, Volume 2: Radar Remote Sensing and Surface Scattering and Emission Theory. Addison-Wesley, London, p 1064

Wismann V (1989) Messung der Windgeschwindigkeit über dem Meer mit einem flugzeugge-tragenen 5.3 GHz Scatterometer. Dissertation zur Erlangung des Grades eines Doktors der Naturwissenschaften, Universität Bremen, Bremen, Germany, S 119

Chapter 4
Doppler Navigation System Application for Measuring Backscattering Signature and Wind Over Water

4.1 Doppler Navigation System

DNS is a self-contained radar system that uses the Doppler effect (Doppler radar) for measuring ground speed and drift angle of aircraft and performs its dead-reckoning navigation (Sosnovskiy and Khaymovich 1987).

The internationally authorized frequency band of 13.25–13.4 GHz has been allocated for the airborne Doppler navigation radar. A center frequency of 13.325 GHz of the band corresponds to a wavelength of 2.25 cm. This frequency presents a good compromise between too low a frequency, resulting in low-velocity sensitivity and large aircraft antenna size and beam widths, and too high a frequency, resulting in excessive absorption and backscattering effects of the atmosphere and precipitation. (Earlier Doppler radars operated in two somewhat lower frequency bands, centered at 8.8–9.8 GHz, respectively. Nowadays, these bands are no longer used for stand-alone Doppler radars.) (Kayton and Fried 1997).

Measuring wind vector and drift angle of aircraft relies on the change of a Doppler frequency of the signal reflected from the underlying surface, depending on a spatial position of an antenna beam. Usually, a DNS antenna has three beams (λ-configuration; beams 1, 2, and 3) or four beams (x-configuration; beams 1, 2, 3, and 4) located in space as presented in Fig. 4.1. An effective antenna beamwidth of the DNS is of $3° - 10°$ (Kolchinskiy et al. 1975). Power reasons (DNS should operate over water as well as over land) and sensitivity of the DNS to velocity influence a choice of a mounting angle for a beam axis in the vertical plane θ_0.

Figure 4.2 shows curves of NRCS of underlying surface versus incidence angle at the frequency band (Ke-band) currently assigned to the Doppler navigation radar (Kayton and Fried 1997). It is clear from the curves that for most types of terrain the NRCS decreases slowly with increase of the beam incidence angle. However, for water surface at its various conditions, the NRCS falls radically as incidence angle increases and assumes different values for various conditions of sea state or water roughness. At the typical Doppler-radar incidence angles of $15° - 30°$ (Kolchinskiy

© The Author(s), under exclusive license to Springer Nature Switzerland AG 2021
A. Nekrasov, *Foundations for Innovative Application of Airborne Radars*,
SpringerBriefs in Earth Sciences, https://doi.org/10.1007/978-3-030-62942-7_4

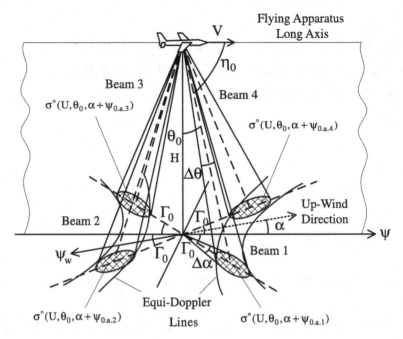

Fig. 4.1 Spatial location of the DNS beams: λ-configuration (*beams 1, 2*, and *3*) and *x*-configuration (*beams 1, 2, 3*, and *4*)

et al. 1975), NRCS is considerably smaller for most sea states than for land, and decreases markedly for a smoother sea state. Therefore, a conservative Doppler-radar design is based on NRCS for the smoothest sea state over which the aircraft is expected to navigate. (Very smooth sea states are relatively rare).

There are two basic antenna system concepts used for measuring drift angle by means of Doppler radar. These are the fixed-antenna system, which is used in most modern systems, and the track-stabilized (roll-and-pitch-stabilized) antenna system. For physically roll-and-pitch-stabilized antenna systems, the value of a current incidence angle remains essentially constant and equal to the chosen design value. For fixed-antenna systems, a conservative design is based on NRCS and range at the largest incidence angle that would be expected for the largest combination of pitch-and-roll angles of the aircraft (Kayton and Fried 1997).

Choice of a mounting angle for a beam axis in the inclined plane η_0 (nominal angle between the antenna longitudinal axis and the beam direction) presents a compromise between high sensitivity to velocity and over-water accuracy, which increases with smaller mounting angles for a beam axis in the inclined plane, and high signal return over water, which increases at larger mounting angles for a beam axis in the inclined plane. Most DNSs use a mounting angle for a beam axis in the inclined plane of somewhere between 65° and 80° (Kayton and Fried 1997). Choice of a mounting angle for a beam axis in the horizontal plane Γ_0 depends on the desired sensitivity to drift, which tends to increase with increasing mounting angle.

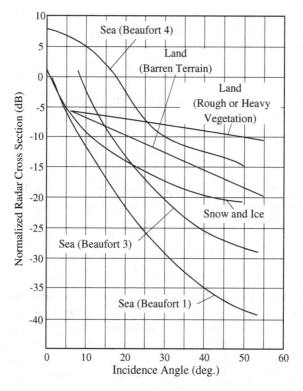

Fig. 4.2 NRCS versus incidence angle for different terrains at Ke-band (Kayton and Fried 1997)

For a typical Doppler radar, mounting angles for a beam axis in the horizontal plane are of $15° - 45°$ (Kolchinskiy et al. 1975).

The relationship between the mounting angles is given by the following equation (Kayton and Fried 1997; Kolchinskiy et al. 1975)

$$\cos \eta_0 = \cos \Gamma_0 \cos \theta_0. \tag{4.1}$$

The mounting angle for a beam axis in the horizontal plane should satisfy the following condition $\Gamma_0 > \beta_{dr.\max}$, where $\beta_{dr.\max}$ is the maximum possible drift angle (Sosnovskiy and Khaymovich 1987). The mounting angle for a beam axis in the inclined plane is defined by requirements to the width of a Doppler spectrum of the reflected signal Δf_D, which depends on the effective antenna beamwidth in the inclined plane $\theta_{a.incl}$; $\theta_{a.incl} \approx 5°$ for the DNS. The relative width of a Doppler spectrum $\Delta f_D/F_D$ is given by Davydov et al. (1977)

$$\frac{\Delta f_D}{F_D} = \frac{\theta_{a.incl}}{\sqrt{2}} \tan \eta_0, \tag{4.2}$$

where F_D is the Doppler frequency, $F_D = \frac{2V_g}{\lambda} \cos \eta_0$, V_g is the aircraft velocity relative to the ground.

To perform high accuracy measurements with the DNS, the following condition should be satisfied (Davydov et al. 1977)

$$\frac{\Delta f_D}{F_D} \leq 0.1 \div 0.2. \tag{4.3}$$

Thus, from (4.2) and (4.3), the mounting angle for a beam axis in the inclined plane should satisfy the following condition

$$\eta_0 \leq \arctan\left[(0.1 \div 0.2)\frac{\sqrt{2}}{\theta_{a.incl}}\right]. \tag{4.4}$$

From (4.4), assuming that the effective antenna beamwidth in the inclined plane is typical and so it equals to 5°, the condition of choice of a mounting angle for a beam axis in the inclined plane is as follows

$$\eta_0 \leq 58.3° \div 72.2°. \tag{4.5}$$

Then, using (4.1), the areas of admissible mounting angles for beam axes could be obtained. Lower limits corresponding to maximum admissible mounting angles for a beam axis in the inclined plane and area of typical mounting angles for a beam axes in the vertical and horizontal planes are presented in Fig. 4.3. Trace $\Delta f_D/F_D = 0.1$ ($\eta_0 = 58.3°$) presents the lower limit of high accuracy of measurement. Trace $\Delta f_D/F_D = 0.2$ ($\eta_0 = 72.9°$) is the lower limit of sufficient high accuracy of measurement. Table 4.1 also gives some combinations of mounting angles of beam axes in the vertical and horizontal planes for two lower limits corresponding to the maximum admissible mounting angles of beam axis in the inclined plane.

The DNS multi-beam antenna allows selecting a power backscattered by the underlying surface from different directions, namely from directions corresponding to the appropriate beam relative to the aircraft course ψ, e.g., $\psi_{0.a.1}$, $\psi_{0.a.2}$, $\psi_{0.a.3}$, and $\psi_{0.a.4}$ as shown in Fig. 4.1. Each beam provides angular resolutions in azimuthal and vertical planes, $\Delta\alpha$ and $\Delta\theta$, respectively. Therefore, those features can be used for DNS enhancement to measure the water-surface backscattering signature and wind vector over water surface when it operates in a scatterometer mode, in addition to its standard navigation application (Nekrasov 2010a, 2012a).

4.2 Measuring Water-Surface Backscattering Signature Using Doppler Navigation System

If an aircraft makes a horizontal rectilinear flight with speed V at some altitude H above mean sea surface, and the DNS has a roll-and-pitch-stabilized antenna system, the NRCS values obtained with beams 1, 2, 3, 4 would be $\sigma°(\theta_0, \psi + \psi_{0.a.1})$, $\sigma°(\theta_0, \psi + \psi_{0.a.2})$, $\sigma°(\theta_0, \psi + \psi_{0.a.3})$, and $\sigma°(\theta_0, \psi + \psi_{0.a.4})$, respectively, where $\psi_{0.a.1} = \Gamma_0$, $\psi_{0.a.2} = 180° - \Gamma_0$, $\psi_{0.a.3} = 180° + \Gamma_0$, $\psi_{0.a.4} = 360° - \Gamma_0$.

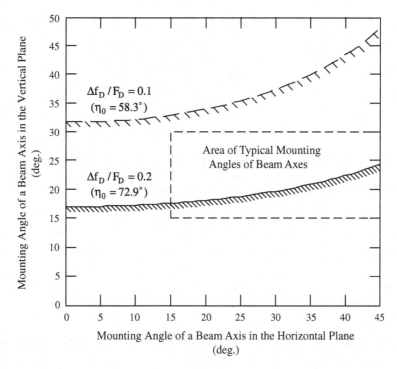

Fig. 4.3 Lower limits corresponding to maximum admissible mounting angles of beam axis in the inclined plane and area of typical mounting angles for beam axes in the *vertical* and *horizontal* planes: trace $\Delta f_D/F_D = 0.1$ $(\eta_0 = 58.3°)$ presents the lower limit of high accuracy of measurement; trace $\Delta f_D/F_D = 0.2$ $(\eta_0 = 72.9°)$ is the lower limit of sufficient high accuracy of measurement; dash line is the contour of the area of typical mounting angles for beam axes in the vertical and horizontal planes

Table 4.1 Combinations of mounting angles for beam axes in vertical and horizontal planes for two lower limits corresponding to the maximum admissible mounting angles for a beam axis in the inclined plane

Γ_0		15°	30°	45°
θ_0	At $\eta_0 = 58.3°$ $(\Delta f_D/F_D = 0.1)$	33°	37.4°	48°
	At $\eta_0 = 72.9°$ $(\Delta f_D/F_D = 0.2)$	17.7°	19.9°	24.6°

As azimuth NRCS is obtained using circle track flight for a scatterometer with an inclined one-beam fixed-position antenna (Masuko et al. 1986), one beam of the DNS operating in the scatterometer mode can be used.

Let beam 1 be used to measure water-surface backscattering signature because both λ- and x-configured DNS have it. As beam 1 axis is directed to the right side of a straight flight, and its typical mounting angle in the vertical plane is not too far from nadir, the anticlockwise circle flight with left roll should be performed to observe the water surface at a medium incidence angle.

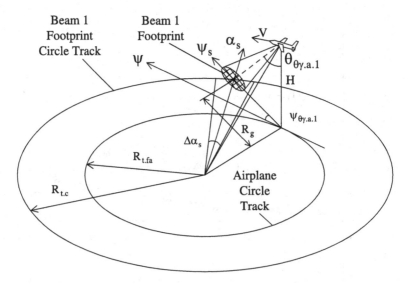

Fig. 4.4 Anticlockwise *circle flight* geometry for measuring the water-surface backscattering signature by the DNS

Let the DNS have a fixed-antenna system (physically non-stabilized to the local horizontal) and the aircraft make a circle flight in accordance with Fig. 4.4 (Nekrasov 2010a).

Therefore, the values of incidence angle of the beam and the beam location in the azimuthal plane will be not equal to the chosen design values, and so an actual incidence angle of beam 1 $\theta_{\theta\gamma.a.1}$ and its actual azimuth direction $\psi_{\theta\gamma.a.1}$ relative to the aircraft current course (aircraft ground track) are as follows (Nekrassov 1998)

$$\theta_{\theta\gamma.a.1} = \arctan\left(\sqrt{\tan^2(\arctan(\tan\theta_0 \sin\psi_{0.a.1}) + \gamma_{fa}) + \rightarrow} \right.$$
$$\left. \overline{\rightarrow \tan^2(\arctan(\tan\theta_0 \cos\psi_{0.a.1}) + \theta_{fa})} \right) \quad (4.6)$$

$$\psi_{\theta\gamma.a.1} = \begin{cases} \arctan\left(\dfrac{\tan(\arctan(\tan\theta_0 \sin\psi_{0.a.1}) + \gamma_{fa})}{\tan(\arctan(\tan\theta_0 \cos\psi_{0.a.1}) + \theta_{fa})}\right) \\ for \ \tan(\arctan(\tan\theta_0 \cos\psi_{0.a.1}) + \theta_{fa}) \geq 0, \\ \pi + \arctan\left(\dfrac{\tan(\arctan(\tan\theta_0 \sin\psi_{0.a.1}) + \gamma_{fa})}{\tan(\arctan(\tan\theta_0 \cos\psi_{0.a.1}) + \theta_{fa})}\right) \\ for \ \tan(\arctan(\tan\theta_0 \cos\psi_{0.a.1}) + \theta_{fa}) < 0, \end{cases} \quad (4.7)$$

where $\theta_{0.a.1}$ is the azimuthal mounting angle for beam 1 axis relative to the aircraft course ψ, $\psi_{0.a.1} = \Gamma_0$, γ_{fa} is the roll angle of flying apparatus (right roll is positive), θ_{fa} is the pitch angle of flying apparatus (pull-up is positive). Then, current NRCS value obtained with beam 1 is $\sigma°(\theta_{\theta\gamma.a.1}, \psi + \psi_{\theta\gamma.a.1})$.

The radius of the aircraft turn is given by (3.3). The ground range and the radius of turn of a selected cell's center point are described by the following expressions obtained using the geometry of Fig. 4.4

$$R_g = \frac{H}{\tan \theta}, \tag{4.8}$$

$$R_{t.c} = \sqrt{R_{t.fa}^2 + R_g^2 + 2R_{t.fa}R_g \sin \psi_{\theta\gamma.a.1}}. \tag{4.9}$$

Time of the aircraft 360° turn is given by (3.5).

Let the middle azimuth of the observed sector be the azimuth of the sector, the azimuth size of the sector relative to the center point of circle of the aircraft track be $\Delta\alpha_s$, and the middle azimuth of the sector be α_s. Then, NRCS samples obtained from the sector and averaged over all measurement values in that sector give NRCS value $\sigma°(\theta_{\theta\gamma.a.1}, \psi_s)$ corresponding to the real observation azimuth angle of the sector ψ_s that is (Nekrasov 2012a)

$$\psi_s = \psi_{\psi_s} + \psi_{\theta\gamma.a.1} \pm 360°, \tag{4.10}$$

where ψ_{ψ_s} is the aircraft course corresponding to the real observation azimuth angle of the sector.

The aircraft performs an anticlockwise turn, and thus, the real observation azimuth angles of the sector start and end are

$$\psi_{s.b} = \psi_s + \Delta\alpha_s/2 \pm 360°, \tag{4.11}$$

$$\psi_{s.e} = \psi_s - \Delta\alpha_s/2 \pm 360°. \tag{4.12}$$

The observation time of a sector and the number of samples that can be obtained from the sector are given by (3.9) and (3.10).

The measured NRCS data set is in fact discrete, and so, each NRCS value obtained is $\sigma°(\theta_{\theta\gamma.a.1}, \psi_{s.i})$, where $i = \overrightarrow{1, N}$, N is the number of sectors observed during aircraft 360° turn at the NRCS measurement, $N = 360°/\Delta\alpha_s$.

Thus, to obtain an azimuth NRCS curve of water surface at medium incidence angles by a DNS operating in scatterometer mode and using a fore-beam directed to the right side at a typical mounting angle in the vertical plane that is not too far from nadir at a straight flight, the measurement should be performed under aircraft anticlockwise circle flight in accordance with a diagram in Fig. 4.5.

The measurement is started when a stable horizontal circle flight at the given altitude, speed of flight, roll and pitch has been established. The measurement is finished when the azimuth of the measurement start is reached. To obtain a greater number of NRCS samples for each sector observed, several consecutive full circle 360° turns may be done.

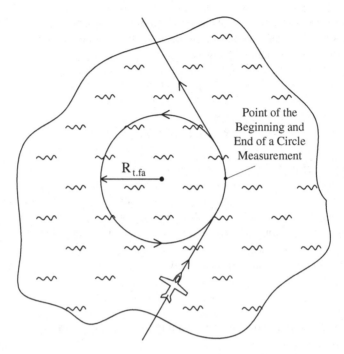

Fig. 4.5 Anticlockwise *circle flight* scheme for measuring water-surface backscattering signature

4.3 Wind Retrieval from Azimuth NRCS Data Obtained with Doppler Navigation System

Let the sea-surface backscattering for medium incidence angles be of the form (2.1). The NRCS model function for medium incidence angles (2.1) can be used without any correction for wind measurement while the azimuth angular size of a cell is below $15° - 20°$ that has been proved by (3.18)–(3.20), Fig. 3.4, and Table 3.2. As the measured NRCS data set is also a function of the wind speed, each NRCS value obtained $\sigma°(\theta_{\theta\gamma.a.1}, \psi_s)$ is considered now as $\sigma°(U, \theta, \psi_{s.i})$.

Let us assume that the angle between the up-wind direction and the first NRCS azimuth $\psi_{s.1}$ is α, and the width of the sector observed is $\Delta\alpha_s$. Then, measured NRCSs $\sigma°(U, \theta, \psi_{s.1})$, $\sigma°(U, \theta, \psi_{s.2})$,..., $\sigma°(U, \theta, \psi_{s.N})$ can be presented as $\sigma°(U, \theta, \alpha)$, $\sigma°(U, \theta, \alpha - \Delta\alpha)$,..., $\sigma°(U, \theta, \alpha - (N - 1)\Delta\alpha)$, respectively, where $i = \overrightarrow{1, N}$, N is the number of sectors forming the $360°$ azimuth NRCS curve, $N = 360°/\Delta\alpha_s$. Taking into account that the anticlockwise circle flight is used to obtain the azimuth NRCS data set, wind speed and up-wind direction can be found from the following system of N equations

$$
\begin{cases}
\sigma^\circ(U, \theta, \alpha) = A(U, \theta) + B(U, \theta) \cos \alpha \\
\qquad + C(U, \theta) \cos(2\alpha), \\
\sigma^\circ(U, \theta, \alpha - \Delta\alpha_s) = A(U, \theta) \\
\qquad + B(U, \theta) \cos(\alpha - \Delta\alpha_s) \\
\qquad + C(U, \theta) \cos(2(\alpha - \Delta\alpha_s)), \\
\cdots\cdots\cdots\cdots\cdots\cdots\cdots\cdots\cdots\cdots\cdots\cdots\cdots\cdots \\
\sigma^\circ(U, \theta, \alpha - (N-2)\Delta\alpha_s) = A(U, \theta) \\
\qquad + B(U, \theta) \cos(\alpha - (N-2)\Delta\alpha_s) \\
\qquad + C(U, \theta) \cos(2(\alpha - (N-2)\Delta\alpha_s)), \\
\sigma^\circ(U, \theta, \alpha - (N-1)\Delta\alpha_s) = A(U, \theta) \\
\qquad + B(U, \theta) \cos(\alpha - (N-1)\Delta\alpha_s) \\
\qquad + C(U, \theta) \cos(2(\alpha - (N-1)\Delta\alpha_s)),
\end{cases}
\tag{4.13}
$$

and then, navigation wind direction can be obtained from (3.22).

The wind vector retrieval procedure (4.13) can be simplified and sped up significantly because an entire 360° azimuth NRCS data set is available. The wind speed can be found from the following equation

$$
U = \left(\frac{\sum_{i=1}^{N} \sigma^\circ(U, \theta, \alpha - (i-1)\Delta\alpha_s)}{N a_0(\theta)} \right)^{1/\gamma_0(\theta)}.
\tag{4.14}
$$

Then, knowing the wind speed from (4.14), the up-wind direction can be estimated using (3.26) and (3.27).

Otherwise, the general formula for the NRCS taking into account the azimuth sector width (3.19) also can be used for that purpose. Let the wide azimuth sector NRCS be $\sigma^\circ(U, \theta, \alpha, \Delta\alpha_w)$, where $\Delta\alpha_w$ is the width of the wide azimuth sector that is much wider than the azimuth size of a sector observed under azimuth NRCS sampling $\Delta\alpha_s$. Then, the 360° NRCS data set obtained should be divided into 4 quadrants (wide azimuth sectors of 90° width) $\alpha + 45° \pm \Delta\alpha_w/2$, $\alpha + 135° \pm \Delta\alpha_w/2$, $\alpha + 225° \pm \Delta\alpha_w/2$, $\alpha + 315° \pm \Delta\alpha_w/2$, and the NRCSs for each quadrant $\sigma^\circ(U, \theta, \alpha + 45°, \Delta\alpha_w)$, $\sigma^\circ(U, \theta, \alpha + 135°, \Delta\alpha_w)$, $\sigma^\circ(U, \theta, \alpha + 225°, \Delta\alpha_w)$, and $\sigma^\circ(U, \theta, \alpha + 315°, \Delta\alpha_w)$, respectively, should be calculated using the following general equation

$$
\sigma^\circ(U, \theta, \alpha, \Delta\alpha_w) = \frac{\Delta\alpha_s}{\Delta\alpha_w} \sum_{i=-\frac{\Delta\alpha_w}{2\Delta\alpha_s}}^{\frac{\Delta\alpha_w}{2\Delta\alpha_s}} \sigma^\circ(U, \theta, \alpha + i\Delta\alpha_s).
\tag{4.15}
$$

Then, using (3.19), two pairs of possible up-wind directions $\alpha_{1,2}$ and $\alpha_{3,4}$ can be obtained as follows

$$
\alpha_{1,2} = \pm \arccos\left(\frac{\sigma^\circ(U, \theta, \alpha + 45°, \Delta\alpha_w) - \sigma^\circ(U, \theta, \alpha + 225°, \Delta\alpha_w)}{2k_1(\Delta\alpha_w)B(U, \theta)} \right) - 45°,
$$

$$
\tag{4.16}
$$

$$\alpha_{3,4} = \pm \arccos \left(\frac{\sigma^\circ(U, \theta, \alpha + 135^\circ, \Delta\alpha_w) - \sigma^\circ(U, \theta, \alpha + 315^\circ, \Delta\alpha_w)}{2k_1(\Delta\alpha_w)B(U, \theta)} \right) - 135^\circ.$$

$$(4.17)$$

The nearest two up-wind directions of those up-wind direction pairs obtained (one from $\alpha_{1,2}$ and one from $\alpha_{3,4}$) will give the estimated true up-wind direction α. Then, wind direction can be found from (3.23).

Thus, estimated wind speed from (4.14) and up-wind direction from (4.16)– (4.17) simplifies solving the system of Eq. (4.13) when they are used as estimates for the wind speed and up-wind direction in (4.13) speeding up the wind vector retrieval procedure significantly.

4.4 Measuring Wind Vector by Doppler Navigation System at a Rectilinear Flight

As the DNS multi-beam antenna allows selecting power backscattered by underlying surface from different directions, namely from directions corresponding to the appropriate beam relative to the aircraft course, as shown in Fig. 4.1, and as three or four NRCS values obtained from considerably different azimuth directions are quite enough to measure wind vector over water by the intensity of the reflected signal (Nekrassov 1997), an airborne DNS also can be used as a multi-beam (three- or four-beam) scatterometer for recovering near-surface wind speed and direction. For this purpose, an airborne DNS having the following mounting angles for antenna beam axes $\theta_0 = 30^\circ$ and $\Gamma_0 = 30^\circ \div 45^\circ$, or $\theta_0 > 30^\circ$ ($\theta_0 \rightarrow 45^\circ$) and $\Gamma_0 = 30^\circ \div 45^\circ$, could be used. The second case requires increased transmitted power in comparison with the first case. Nevertheless, it permits better usage of anisotropic properties of water-surface scattering at medium incidence angles for measuring the near-surface wind vector as well as for increasing accuracy of measurement for typical DNS parameters (Nekrasov 2010b).

As a DNS can have a fixed-antenna system or a track-stabilized (roll-and-pitch-stabilized) antenna system with three or four beams, these constructive features of DNS should be taken into account when developing a measuring algorithm for the wind vector over water.

Let a flying apparatus equipped with a DNS perform a horizontal rectilinear flight with the speed V at some altitude H above mean sea surface the DNS use a roll-and-pitch-stabilized antenna (physically stabilized to the local horizontal), and thus, the value of incidence angle θ remains essentially constant and equal to the chosen design value θ_0. Let the aircraft velocity vector V be directed along the intersection of the local horizontal plane and the local vertical plane through the longitudinal axis of aircraft (condition for no-drift angle and no-climb angle) which means that the aircraft flight is horizontal and the aircraft course ψ is the same as the aircraft ground track. The directions of DNS beams 1, 2, 3, and 4 relative to the aircraft course are $\psi_{0.a.1}$, $\psi_{0.a.2}$, $\psi_{0.a.3}$, and $\psi_{0.a.4}$, respectively (Fig. 4.1). Let sea surface wind blow in direction ψ_w, and the angle between the up-wind direction and the aircraft course is α. Let the NRCS model function for medium incidence angles be of the form (2.1). Then, the NRCS values obtained with beams 1, 2, 3, and 4 are $\sigma°(U, \theta_0, \alpha + \psi_{0.a.1})$, $\sigma°(U, \theta_0, \alpha + \psi_{0.a.2})$, $\sigma°(U, \theta_0, \alpha + \psi_{0.a.3})$,   and $\sigma°(U, \theta_0, \alpha + \psi_{0.a.4})$, respectively.

In the general case, the wind speed and up-wind direction can be found from a system of three equations in case of a three-beam DNS

$$
\begin{cases}
\sigma°(U, \theta_0, \alpha + \psi_{0.a.1}) = A(U, \theta_0) \\
\qquad + B(U, \theta_0) \cos(\alpha + \psi_{0.a.1}) \\
\qquad + C(U, \theta_0) \cos(2(\alpha + \psi_{0.a.1})), \\
\sigma°(U, \theta_0, \alpha + \psi_{0.a.2}) = A(U, \theta_0) \\
\qquad + B(U, \theta_0) \cos(\alpha + \psi_{0.a.2}) \\
\qquad + C(U, \theta_0) \cos(2(\alpha + \psi_{0.a.2})), \\
\sigma°(U, \theta_0, \alpha + \psi_{0.a.3}) = A(U, \theta_0) \\
\qquad + B(U, \theta_0) \cos(\alpha + \psi_{0.a.3}) \\
\qquad + C(U, \theta_{0.a.3}) \cos(2(\alpha + \psi_{0.a.3})),
\end{cases}
\tag{4.18}
$$

or from a system of four equations in case of the four-beam DNS

$$
\begin{cases}
\sigma°(U, \theta_0, \alpha + \psi_{0.a.1}) = A(U, \theta_0) \\
\qquad + B(U, \theta_0) \cos(\alpha + \psi_{0.a.1}) \\
\qquad + C(U, \theta_0) \cos(2(\alpha + \psi_{0.a.1})), \\
\sigma°(U, \theta_0, \alpha + \psi_{0.a.2}) = A(U, \theta_0) \\
\qquad + B(U, \theta_0) \cos(\alpha + \psi_{0.a.2}) \\
\qquad + C(U, \theta_0) \cos(2(\alpha + \psi_{0.a.2})), \\
\sigma°(U, \theta_0, \alpha + \psi_{0.a.3}) = A(U, \theta_0) \\
\qquad + B(U, \theta_0) \cos(\alpha + \psi_{0.a.3}) \\
\qquad + C(U, \theta_{0.a.3}) \cos(2(\alpha + \psi_{0.a.3})), \\
\sigma°(U, \theta_0, \alpha + \psi_{0.a.4}) = A(U, \theta_0) \\
\qquad + B(U, \theta_0) \cos(\alpha + \psi_{0.a.4}) \\
\qquad + C(U, \theta_0) \cos(2(\alpha + \psi_{0.a.4})).
\end{cases}
\tag{4.19}
$$

Then, the wind direction can be found from the following equation

$$
\psi_w = \psi - \alpha \pm 180°.
\tag{4.20}
$$

A special case for retrieval of the wind vector with a four-beam DNS having the roll-and-pitch-stabilized antenna is when a mounting angle for a beam axis in the horizontal plane is equal to 45°. In that case, the NRCS values obtained with beams 1, 2, 3, and 4 are $\sigma°(U, \theta_0, \alpha + 45°)$, $\sigma°(U, \theta_0, \alpha + 135°)$, $\sigma°(U, \theta, \alpha + 225°)$, and $\sigma°(U, \theta_0, \alpha + 315°)$. Then, the following algorithm to estimate the wind vector over the sea surface can be proposed.

The wind speed can be found from the following equation (Nekrasov 2005)

$$U = \left(\frac{A(U, \theta_0)}{a_0(\theta_0)}\right)^{1/\gamma_0(\theta_0)}$$

$$= \left(\frac{\begin{matrix}\sigma°(U, \theta_0, \alpha + 45°) + \sigma°(U, \theta_0, \alpha + 135°) \\ \rightarrow\ + \sigma°(U, \theta_0, \alpha + 225°) + \sigma°(U, \theta_0, \alpha + 315°)\end{matrix}}{4a_0(\theta_0)}\right)^{1/\gamma_0(\theta_0)} \tag{4.21}$$

To find the wind direction, at first, the space of possible solutions could be divided into four quadrants. Then, a quadrant containing the solution should be found, and the unique wind direction can be obtained as follows

$$\begin{aligned} &\text{if } \sigma°(U, \theta_0, \alpha + 45°) < \sigma°(U, \theta_0, \alpha + 225°) \\ &\text{and } \sigma°(U, \theta_0, \alpha + 135°) \geq \sigma°(U, \theta_0, \alpha + 315°) \\ &\qquad \Rightarrow \psi_w = \psi + 45° + \alpha_q \pm 180°, \\ &\text{if } \sigma°(U, \theta_0, \alpha + 45°) \leq \sigma°(U, \theta_0, \alpha + 225°) \\ &\text{and } \sigma°(U, \theta_0, \alpha + 135°) < \sigma°(U, \theta_0, \alpha + 315°) \\ &\qquad \Rightarrow \psi_w = \psi + 225° - \alpha_q \pm 180°, \\ &\text{if } \sigma°(U, \theta_0, \alpha + 45°) > \sigma°(U, \theta_0, \alpha + 225°) \\ &\text{and } \sigma°(U, \theta_0, \alpha + 135°) \leq \sigma°(U, \theta_0, \alpha + 315°) \\ &\qquad \Rightarrow \psi_w = \psi + 225° + \alpha_q \pm 180°, \\ &\text{if } \sigma°(U, \theta_0, \alpha + 45°) \geq \sigma°(U, \theta_0, \alpha + 225°) \\ &\text{and } \sigma°(U, \theta_0, \alpha + 135°) > \sigma°(U, \theta_0, \alpha + 315°) \\ &\qquad \Rightarrow \psi_w = \psi + 45° - \alpha_q \pm 180°, \end{aligned} \tag{4.22}$$

where α_q is the angle of the wind in the quadrant,

$$\alpha_q = \begin{cases} 0°, & A_1 \geq 1 \\ 0.5 \arccos A_1, & -1 < A_1 < 1, \\ 90°, & A_1 \leq 1 \end{cases} \tag{4.23}$$

$$A_1 = \frac{\sigma°(U, \theta_0, \alpha + 45°) + \sigma°(U, \theta_0, \alpha + 225°) - 2A(U, \theta_0)}{2C(U, \theta_0)}. \tag{4.24}$$

Now, let a flying apparatus equipped with a DNS having a multi-beam fixed-antenna system (physically non-stabilized to the local horizontal) make a horizontal rectilinear flight with the speed V at some altitude H above the mean sea surface. As the antenna system is not stabilized to the local horizontal, the values of incidence angles of beams and beam locations in the azimuthal plane are not equal to the chosen design values. Thus, the actual incidence angle of beam N $\theta_{\theta\gamma.a.N}$ and the actual azimuth direction of beam N $\psi_{\theta\gamma.a.N}$ relative to the aircraft course (aircraft ground track) can be found from (4.6) and (4.7) by substitution of 1 in subscripts by the appropriate beam number, and the NRCS values obtained with beams 1, 2, 3, and 4 are $\sigma°(U, \theta_{\theta\gamma.a.1}, \alpha + \psi_{\theta\gamma.a.1})$, $\sigma°(U, \theta_{\theta\gamma.a.2}, \alpha + \psi_{\theta\gamma.a.2})$, $\sigma°(U, \theta_{\theta\gamma.a.3}, \alpha + \psi_{\theta\gamma.a.3})$, and $\sigma°(U, \theta_{\theta\gamma.a.4}, \alpha + \psi_{\theta\gamma.a.4})$, respectively.

As the DNS considered has the antenna system that is physically non-stabilized to a local horizontal plane, the difference between those actual angles and the mounting angles can be significant even at horizontal flight. For example, five-degree roll-and-pitch combinations at a mounting angle for a beam axis in the vertical plane of 30° and an arbitrary mounting angle for a beam axis in the azimuthal plane may lead to a beam axis shift up to 6.4° in the vertical plane and up to 14.4° in the azimuthal plane. The same roll-and-pitch combinations at a mounting angle for a beam axis in the vertical plane of 45° lead to a lesser beam axis shift up to 5.5° in the vertical plane and up to 10.6° in the azimuthal plane. So, it is desirable that the mounting angle for a beam axis in the vertical plane tends to 45°. Also, such possible differences of the actual beam axis angles from the mounting angles should be taken into account under the development of the measuring algorithm (Nekrasov 2008a).

Thus, wind speed and up-wind direction can be found from a system of three equations in case of the three-beam DNS (Nekrasov 2008b)

$$
\begin{cases}
\sigma°(U, \theta_{\theta\gamma.a.1}, \alpha + \psi_{\theta\gamma.a.1}) = A(U, \theta_{\theta\gamma.a.1}) \\
\qquad\qquad + B(U, \theta_{\theta\gamma.a.1})\cos(\alpha + \psi_{\theta\gamma.a.1}) \\
\qquad\qquad + C(U, \theta_{\theta\gamma.a.1})\cos(2(\alpha + \psi_{\theta\gamma.a.1})), \\
\sigma°(U, \theta_{\theta\gamma.a.2}, \alpha + \psi_{\theta\gamma.a.2}) = A(U, \theta_{\theta\gamma.a.2}) \\
\qquad\qquad + B(U, \theta_{\theta\gamma.a.2})\cos(\alpha + \psi_{\theta\gamma.a.2}) \\
\qquad\qquad + C(U, \theta_{\theta\gamma.a.2})\cos(2(\alpha + \psi_{\theta\gamma.a.2})), \\
\sigma°(U, \theta_{\theta\gamma.a.3}, \alpha + \psi_{\theta\gamma.a.3}) = A(U, \theta_{\theta\gamma.a.3}) \\
\qquad\qquad + B(U, \theta_{\theta\gamma.a.3})\cos(\alpha + \psi_{\theta\gamma.a.3}) \\
\qquad\qquad\; C(U, \theta_{\theta\gamma.a.3})\cos(2(\alpha + \psi_{\theta\gamma.a.3})).
\end{cases}
\tag{4.25}
$$

or from a system of four equations in case of the four-beam DNS (Nekrasov 2008b)

$$
\begin{cases}
\sigma^\circ(U, \theta_{\theta\gamma.a.1}, \alpha + \psi_{\theta\gamma.a.1}) = A(U, \theta_{\theta\gamma.a.1}) \\
\qquad\qquad + B(U, \theta_{\theta\gamma.a.1})\cos(\alpha + \psi_{\theta\gamma.a.1}) \\
\qquad\qquad + C(U, \theta_{\theta\gamma.a.1})\cos(2(\alpha + \psi_{\theta\gamma.a.1})), \\
\sigma^\circ(U, \theta_{\theta\gamma.a.2}, \alpha + \psi_{\theta\gamma.a.2}) = A(U, \theta_{\theta\gamma.a.2}) \\
\qquad\qquad + B(U, \theta_{\theta\gamma.a.2})\cos(\alpha + \psi_{\theta\gamma.a.2}) \\
\qquad\qquad + C(U, \theta_{\theta\gamma.a.2})\cos(2(\alpha + \psi_{\theta\gamma.a.2})), \\
\sigma^\circ(U, \theta_{\theta\gamma.a.3}, \alpha + \psi_{\theta\gamma.a.3}) = A(U, \theta_{\theta\gamma.a.3}) \\
\qquad\qquad + B(U, \theta_{\theta\gamma.a.3})\cos(\alpha + \psi_{\theta\gamma.a.3}) \\
\qquad\qquad + C(U, \theta_{\theta\gamma.a.3})\cos(2(\alpha + \psi_{\theta\gamma.a.3})), \\
\sigma^\circ(U, \theta_{\theta\gamma.a.4}, \alpha + \psi_{\theta\gamma.a.4}) = A(U, \theta_{\theta\gamma.a.4}) \\
\qquad\qquad + B(U, \theta_{\theta\gamma.a.4})\cos(\alpha + \psi_{\theta\gamma.a.4}) \\
\qquad\qquad + C(U, \theta_{\theta\gamma.a.4})\cos(2(\alpha + \psi_{\theta\gamma.a.4})).
\end{cases} \tag{4.26}
$$

The wind direction can be found from (4.20).

4.5 Simulation of Wind Vector Retrieval by Doppler Navigation System at a Rectilinear Flight

To verify the feasibility of DNS with the x-configured four-beam measuring geometries, and algorithms for retrieval of the wind speed and direction over sea, two cases of simulation have been performed. The mounting angles of a beam axis in the horizontal plane have been considered to be equal to 45°, 30°, and 15°. The mounting angles of a beam axis in the vertical plane have been considered to be equal to 30° (the highest DNS mounting angle of a beam axis in the vertical plane from the range of its typical angles) and 45° (higher and not typical mounting angle of a beam axis in the vertical plane).

The first case of simulation is for DNS having the track-stabilized antenna system (Nekrasov et al. 2017a). The second case of simulation has been performed for DNS with the fixed-antenna system (Nekrasov et al. 2017b). A Ku-band geophysical model function of the form of Eq. (2.1) for the horizontal transmit and receive polarization from (Moore and Fung 1979) has been used for this simulation. The "measured" azimuth NRCSs were generated using a Rayleigh Power (Exponential) distribution. Typically, DNS provides a good signal-to-noise ratio making the noise influence insignificant for measurement. Nevertheless, the noise may affect wind retrieval errors at lower NRCS values (lower power of the backscattered signal from the water surface at lower wind speeds). So it is also taken into account under the simulations.

In case of the DNS with the track-stabilized antenna system, a Monte Carlo method with 50 trials at wind speeds of 2–20 m/s have been used to simulate the wind retrieval by solving the appropriate system of Eq. (4.19). Search resolutions in this system of equations were 0.01 m/s and 0.1° for the wind speed and direction, respectively. The instrumental noise of 0.1 dB and 0.2 dB have been taken into account at the mounting angle in the vertical plane of 30° and 45°, respectively. 5000 and 1565 "measured" NRCS samples have been averaged for each azimuthal angle at simulations for the incidence angles of 30° and 45°, respectively. So, two series of simulations have been performed.

The simulation results of the first series at the incidence angles of 30° with the assumption of 0.1 dB instrumental noise for mounting angles of a beam axis in the horizontal plane of 45°, 30°, and 15° are presented in Figs. 4.6, 4.7, and 4.8, respectively.

At the incidence angle of 30°, the simulation with the assumption of 0.1 dB instrumental noise has shown that the maximum errors of the wind speed and direction retrieval for mounting angles of a beam axis in the horizontal plane of 45°, 30°, and 15° are 0.39 m/s and 14.0°, 0.97 m/s and 14.6°, 2.4 m/s and 19.6°, respectively. The wind direction errors are in the typical range of the scatterometer wind direction retrieval error of 20°. Unfortunately, at the worst case of the mounting angle of a beam axis in the horizontal plane of 15°, the wind speed retrieval error is over the typical scatterometer wind retrieval error of 2 m/s. The

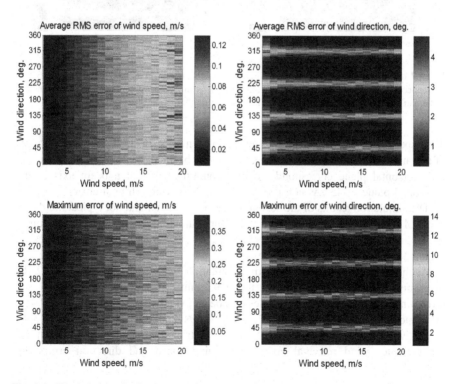

Fig. 4.6 Simulation results for the mounting angle of a beam axis in the horizontal plane of 45° at wind speeds of 2–20 m/s for the incidence angle of 30° with the assumption of 0.1 dB instrumental noise and 5000 averaged NRCS samples for each azimuthal angle

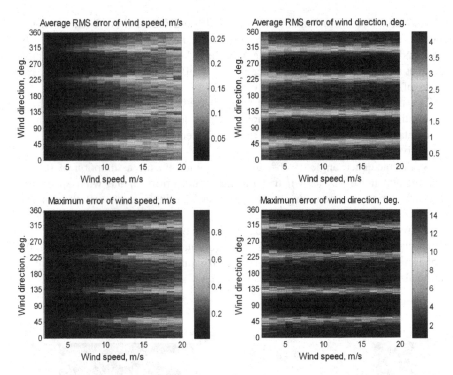

Fig. 4.7 Simulation results for the mounting angle of a beam axis in the horizontal plane of 30° at wind speeds of 2–20 m/s for the incidence angle of 30° with the assumption of 0.1 dB instrumental noise and 5000 averaged NRCS samples for each azimuthal angle

wind speed and direction errors are increasing with decreasing the mounting angle of a beam axis in the horizontal plane. Hence, the best location of the DNS beams in the horizontal plane corresponds to the mounting angle of a beam axis in the horizontal plane of 45°, which provides the lowest wind retrieval errors.

The simulation results of the second series at the higher and not typical incidence angle of 45° with the assumption of 0.2 dB instrumental noise for mounting angles of a beam axis in the horizontal plane of 45°, 30°, and 15° are presented in Figs. 4.9, 4.10, and 4.11, respectively.

At the incidence angle of 45°, the simulation with the assumption of 0.2 dB instrumental noise has shown that the maximum errors of the wind speed and direction retrieval for mounting angles of a beam axis in the horizontal plane of 45°, 30°, and 15° are 0.58 m/s and 7.7°, 0.74 m/s and 8.9°, 1.8 m/s and 13.9°, respectively. All the wind speed and direction retrieval errors are within the typical scatterometer wind retrieval errors. The wind speed and direction errors are

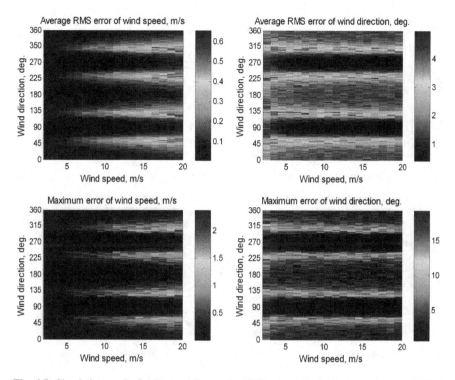

Fig. 4.8 Simulation results for the mounting angle of a beam axis in the horizontal plane of 15° at wind speeds of 2–20 m/s for the incidence angle of 30° with the assumption of 0.1 dB instrumental noise and 5000 averaged NRCS samples for each azimuthal angle

increasing with decreasing the mounting angle of a beam axis in the horizontal plane. As in the previous case considered, the best location of the DNS beams in the horizontal plane corresponds to the mounting angle of a beam axis in the horizontal plane of 45°, which provides the lowest wind retrieval errors.

These examples clearly demonstrate the suitability of the four-beam DNS with the track-stabilized antenna system for the sea wind measurement at the rectilinear flight, and advantage of the 45° mounting angle of a beam axis in the vertical plane in comparison with 30° as well as advantage of the 45° mounting angle of a beam axis in the horizontal plane in comparison with 30°, and especially with 15°.

In case of the DNS with the fixed-antenna antenna system, two series of simulations have been performed at the incidence angles of 30° and 45°. To simulate the wind retrieval, the system of Eq. (4.26) has been used. The simulation at the highest and lowest angles from the typical range of mounting angles of the antenna beam axis in the horizontal plane of 45° and 15° has been performed at the worst case of a cross-wind horizontal rectilinear flight with the angle of attack of −5° and wind speed of 2 m/s. At the incidence angle of 30°, the combinations for $\Gamma_0 = 45°$ are $(\psi_{\vartheta\gamma.a.4} = 307°$ and $\psi_{\vartheta\gamma.a.1} = 53°, \vartheta_{\vartheta\gamma.a.4} = \vartheta_{\vartheta\gamma.a.1} = \vartheta_1 = 27°)$ and $(\psi_{\vartheta\gamma.a.2} =$

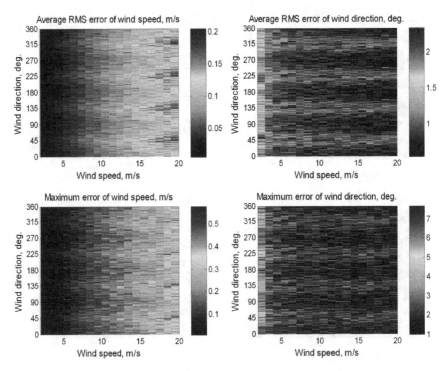

Fig. 4.9 Simulation results for the mounting angle of a beam axis in the horizontal plane of 45° at wind speeds of 2–20 m/s for the incidence angle of 45° with the assumption of 0.2 dB instrumental noise and 1565 averaged NRCS samples for each azimuthal angle

142° and $\psi_{\vartheta\gamma.a.3} = 218°, \vartheta_{\vartheta\gamma.a.2} = \vartheta_{\vartheta\gamma.a.3} = \vartheta_2 = 33°$); for $\Gamma_0 = 15°$ are ($\psi_{\vartheta\gamma.a.4} = 276°$ and $\psi_{\vartheta\gamma.a.1} = 84°, \vartheta_{\vartheta\gamma.a.4} = \vartheta_{\vartheta\gamma.a.1} = \vartheta_1 = 29°$)   and   ($\psi_{\vartheta\gamma.a.2} = 113°$ and $\psi_{\vartheta\gamma.a.3} = 247°, \vartheta_{\vartheta\gamma.a.2} = \vartheta_{b.3} = \vartheta_2 = 31°$). At the incidence angle of 45°, the combinations for 45° are ($\psi_{\vartheta\gamma.a.4} = 310°$ and $\psi_{\vartheta\gamma.a.1} = 50°, \vartheta_{\vartheta\gamma.a.4} = \vartheta_{\vartheta\gamma.a.1} = \vartheta_1 = 43°$)   and   ($\psi_{\vartheta\gamma.a.2} = 140°$ and $\psi_{\vartheta\gamma.a.3} = 220°, \vartheta_{\vartheta\gamma.a.2} = \vartheta_{\vartheta\gamma.a.3} = \vartheta_2 = 48°$);   for $\Gamma_0 = 15°$ are ($\psi_{\vartheta\gamma.a.4} = 280°$ and $\psi_{\vartheta\gamma.a.1} = 80°, \vartheta_{\vartheta\gamma.a.4} = \vartheta_{\vartheta\gamma.a.1} = \vartheta_1 = 44°$) and ($\psi_{\vartheta\gamma.a.2} = 110°$ and $\psi_{\vartheta\gamma.a.3} = 250°, \vartheta_{\vartheta\gamma.a.2} = \vartheta_{\vartheta\gamma.a.3} = \vartheta_2 = 46°$). 1565 "measured" NRCS samples have been averaged for each azimuthal angle and the instrumental noise of 0.2 dB have been taken into account at simulations for both incidence angles.

The simulation results of the first series at the incidence angles of 30° with the assumption of 0.2 dB instrumental noise for mounting angles of a beam axis in the horizontal plane of 45° and 15° are presented in Figs. 4.12, and 4.13, respectively. The NRCS curves by the geophysical model function of the form (2.1) from (Moore and Fung 1979) at the "true" wind speed are shown by solid curves. Crosses and

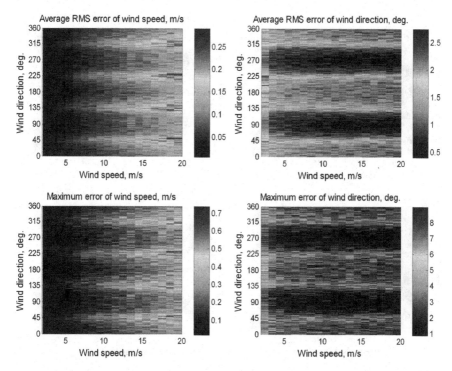

Fig. 4.10 Simulation results for the mounting angle of a beam axis in the horizontal plane of 30° at wind speeds of 2–20 m/s for the incidence angle of 45° with the assumption of 0.2 dB instrumental noise and 1565 averaged NRCS samples for each azimuthal angle

dotted traces represent the "measured" NRCS obtained by integrating 1565 samples for each azimuthal angle with the step of one degree at the actual beam incidence angles of beams θ1 and θ2, respectively. Dashed traces demonstrate the azimuth NRCS curves accordingly to the geophysical model function of the form (2.1) which correspond to the "measured" up-wind directions and wind speeds.

The simulation results of the second series at the incidence angles of 45° with the assumption of 0.2 dB instrumental noise for mounting angles of a beam axis in the horizontal plane of 45° and 15° are presented in Figs. 4.14, and 4.15, respectively.

These examples clearly indicate the suitability of the four-beam DNS with the fixed-antenna antenna system that is not stabilized physically to the local horizontal for the measurement of the wind over the sea at the typical mounting angle of the antenna beam axis in the vertical plane of 30° or higher. The accuracy of the algorithm proposed, even in the considered worst case scenario of 2 m/s wind speed

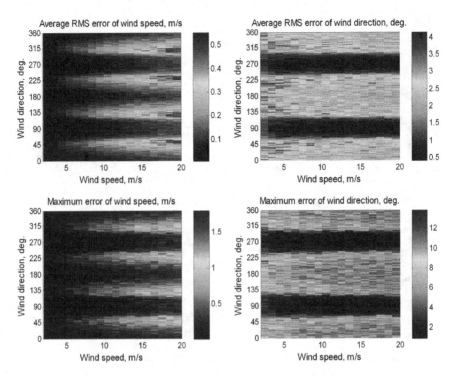

Fig. 4.11 Simulation results for the mounting angle of a beam axis in the horizontal plane of 15° at wind speeds of 2–20 m/s for the incidence angle of 45° with the assumption of 0.2 dB instrumental noise and 1565 averaged NRCS samples for each azimuthal angle

and the typical mounting angle of the antenna beam axis in the horizontal plane of 15°, is within the usual accuracy range for scatterometer wind measurement. These results also indicate that DNS with an increased mounting angle of 45° for the antenna beam axis in the vertical plane provides better usage of anisotropic properties of the sea surface scattering at wind measurements over water.

The altitude applicability of the method considered depends on the DNS beam geometry. Assuming the wind and wave conditions to be identical in the area with a side that does not exceed 15–20 km, the maximum altitude for the wind measurement at the mounting angle of the antenna beam axis in the vertical plane of 30° will be about 24 and 67 km at the mounting angles of the antenna beam axis in the horizontal plane of 45° and 15°, respectively (Fig. 4.16). At the same time, the maximum altitude at the mounting angle of the antenna beam axis in the vertical plane of 45° will be lower in 1.73 times than at 30°.

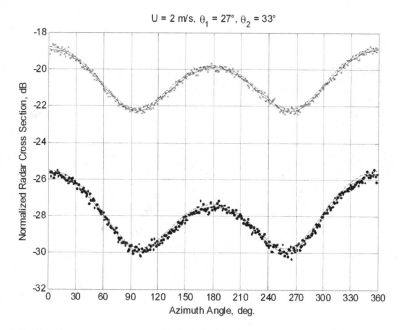

Fig. 4.12 Simulation results for the mounting angle of a beam axis in the horizontal plane of 45° at the "true" wind speed of 2 m/s for the incidence angle of 30° and the angle of attack of −5° with the assumption of 0.2 dB instrumental noise and 1565 averaged NRCS samples for each azimuthal angle: "measured" wind speed of 2 m/s and up-wind direction of 1.2°

4.6   Conclusions to Doppler Navigation System Application for Measuring Backscattering Signature and Wind Over Water

Completed analysis of a Doppler navigation system has shown that the system operating in a scatterometer mode can be used for remote measurements of the sea-surface backscattering signature with a circular flight pattern along with recovering the near-surface wind vector over water from NRCS azimuth curves obtained, as well as for measuring wind vector during a rectilinear flight in addition to its typical navigation application.

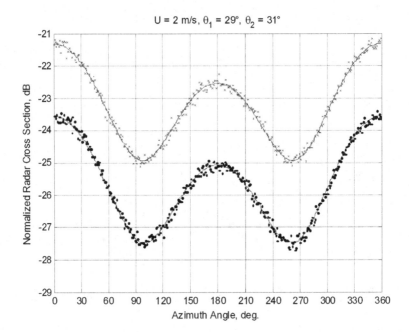

Fig. 4.13 Simulation results for the mounting angle of a beam axis in the horizontal plane of 15° at the "true" wind speed of 2 m/s for the incidence angle of 30° and the angle of attack of −5° with the assumption of 0.2 dB instrumental noise and 1565 averaged NRCS samples for each azimuthal angle: "measured" wind speed of 2.01 m/s and up-wind direction of 1.7°

As the azimuth NRCS curve can be obtained by a scatterometer with an inclined one-beam fixed-position antenna using the circle track flight, one beam of the DNS can be used. The beam should be pointed to the outer side of the aircraft turn to observe a greater area of the water surface and to obtain a greater number of independent NRCS samples. Since the mounting angle of the beam axis in the vertical plane is located not far from nadir (during a straight flight), a circular flight with small roll should be carried out to provide the azimuth NRCS curve measurement and near-surface wind vector estimation in the range of medium incidence angles.

An airborne DNS having the three- or four-beam fixed or roll-and-pitch-stabilized antenna system employed as a multi-beam scatterometer can also be used for measuring the wind vector over water during a rectilinear flight. For this purpose, a DNS should have the following mounting angles for antenna beam axes $\theta_0 = 30°$ and $\Gamma_0 = 30° \div 45°$, or $\theta_0 > 30°$ ($\theta_0 \rightarrow 45°$) and $\Gamma_0 = 30° \div 45°$. The second case requires an increased transmitted power in comparison with the

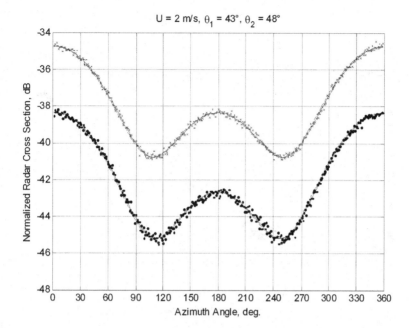

Fig. 4.14 Simulation results for the mounting angle of a beam axis in the horizontal plane of 45° at the "true" wind speed of 2 m/s for the incidence angle of 45° and the angle of attack of −5° with the assumption of 0.2 dB instrumental noise and 1565 averaged NRCS samples for each azimuthal angle: "measured" wind speed of 2 m/s and up-wind direction of 358.7°

first case. Nevertheless, it allows for better use of the anisotropic properties of water-surface scattering at medium incidence angles to achieve measurement of near-surface wind vector, to increase the accuracy of measurement of typical DNS parameters as well as to decrease beam axis deviation due to roll-and-pitch influence, especially for the DNS with a fixed-antenna system.

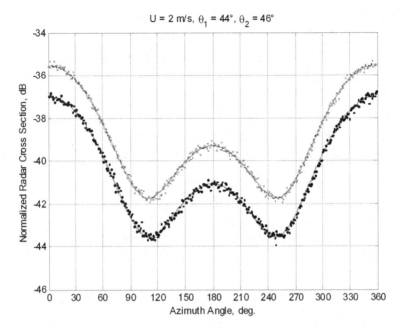

Fig. 4.15 Simulation results for the mounting angle of a beam axis in the horizontal plane of 15° at the "true" wind speed of 2 m/s for the incidence angle of 45° and the angle of attack of −5° with the assumption of 0.2 dB instrumental noise and 1565 averaged NRCS samples for each azimuthal angle: "measured" wind speed of 2 m/s and up-wind direction of 358.9°

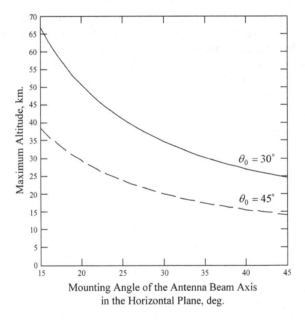

Fig. 4.16 Maximum altitude of DNS wind retrieval algorithm applicability versus mounting angle of the antenna beam axis in the horizontal plane

References

Davydov PS, Zhavoronkov VP, Kashcheyev GV, Krinitsyn VV, Uvarov VS, Khresin IN (1977) Radar systems of flying apparatuses. Transport, Moscow, USSR, p 352 (in Russian)

Kayton M, Fried WR (1997) Avionics navigation systems. Wiley, New York, p 773

Kolchinskiy VYe, Mandurovskiy IA, Konstantinovskiy MI (1975) Autonomous Doppler facilities and systems for navigation of flying apparatus. Sovetskoye Radio, Moscow, USSR, p 432 (in Russian)

Masuko H, Okamoto K, Shimada M, Niwa S (1986) Measurement of microwave backscattering signatures of the ocean surface using X band and Ka band airborne scatterometers. J Geophys Res 91(C11):13065–13083

Moore RK, Fung AK (1979) Radar determination of winds at sea. Proc IEEE 67(11):1504–1521

Nekrasov A (2005) On possibility to measure the sea surface wind vector by the Doppler navigation system of flying apparatus. Proceedings of RADAR 2005, Arlington, Virginia, USA, 9–12 May 2005, pp 747–752

Nekrasov A (2008a) Measuring the sea surface wind vector by the Doppler navigation system of flying apparatus that has a four-beam fixed-antenna system. Proceedings of RADAR 2008, Adelaide, Australia, 2–5 Sep 2008, pp 493–498

Nekrasov A (2008b) Measurement of the sea surface wind vector using an airborne Doppler navigation system with a fixed-antenna system. Materialien zum wissenschaftlichen Seminar der Stipendiaten des "Michail Lomonosov"-Programms 2007/08, Moskau, 18–19 Apr 2008, pp 135–137

Nekrasov A (2010a) Airborne Doppler navigation system application for measurement of the water surface backscattering signature. In: Wagner W, Székely B (eds) ISPRS TC VII symposium – 100 years ISPRS, Vienna, Austria, 2–4 Jul 2010, international archives of the photogrammetry, remote sensing and spatial information sciences, 2010, vol XXXVIII, part 7A, pp 163–168

Nekrasov A (2010b) Microwave measurement of the wind vector over sea by airborne radars. In: Mukherjee M (ed) Advanced microwave and millimeter wave technologies: Semiconductor devices, circuits and systems. In-Tech, Vukovar, pp 521–548

Nekrasov A (2012a) On airborne Doppler navigation system application for measurement of the water surface backscattering signature and estimation of the near-surface wind. Open Rem Sens J 5:15–20

Nekrassov A (1997) Measurement of sea surface wind speed and its navigational direction from flying apparatus. In: Proceedings of Oceans'97, Halifax, Nova Scotia, Canada, 6–9 Oct 1997, pp 83–86

Nekrassov A (1998) Decrease of observation time and influence of flying apparatus list-tangage instability in measurement of sea surface wind vector. In: Proceedings of 5th international conference on remote sensing for marine and coastal environments, San Diego, California, USA, 5–7 Oct 1998, vol 2, pp 514–519

Nekrasov A, Khachaturian A, Gamcová M, Kurdel P, Obukhovets V, Veremyev V, Bogachev M (2017a) Sea wind measurement by Doppler navigation system with x-configured beams in rectilinear flight. Remote Sens 9(9), 889:1–19

Nekrasov A, Khachaturian A, Veremyev V, Bogachev M (2017b) Doppler navigation system with a non-stabilized antenna as a sea-surface wind sensor. Sensors 17(6), 1340:1–10

Sosnovskiy AA, Khaymovich IA (1987) Radio-electronic equipment of flying apparatuses. Transport, Moscow, USSR, p 256 (in Russian)

Chapter 5
Measuring Water-Surface Backscattering Signature and Wind by Means of Airborne Weather Radar

5.1 Airborne Weather Radar

AWR is the type of radar equipment mounted on an aircraft for purposes of weather observation and avoidance, aircraft position finding relative to landmarks, and drift angle measurement (Sosnovsky et al. 1990). The AWR is necessary equipment for any civil airplane. It also must be installed on all civil airliners. Military transport aircrafts are usually equipped by weather radars as well. Due to the specificity of airborne applications, designers of avionics systems always try to use the most efficient progressive methods and reliable engineering solutions that provide flight safety and flight regularity in harsh environments (Yanovsky 2005).

The development of the AWR is mainly associated with growing functionalities in detection of different dangerous weather phenomena. The radar observations performed in weather mode are magnitude detection of reflections from clouds and precipitation and Doppler measurements of the motion of particles within a weather formation. Magnitude detection allows determination of particle type (rain, snow, hail, etc.) and precipitation rate. Doppler measurements can be made to yield estimates of turbulence intensity and wind speed. Reliable determination of the presence and severity of the phenomenon known as wind shear is an important area of study as well (Kayton and Fried 1997).

Nevertheless, the second important assignment of the AWR is providing a pilot with navigation information using earth surface mapping. In this case, a possibility to extract some navigation information that allows determining aircraft position with respect to a geographic map is very important for air navigation. Landmark's coordinates relative to the aircraft that are measured by the AWR give a possibility to set flight computer for more precise and more efficient fulfillment of en-route flight, cargo delivery, and cargo throw down to a given point. These improve tactical possibilities of transport aircraft, airplanes of search-and-rescue service, and local airways (Yanovsky 2005).

© The Author(s), under exclusive license to Springer Nature Switzerland AG 2021   61
A. Nekrasov, *Foundations for Innovative Application of Airborne Radars*,
SpringerBriefs in Earth Sciences, https://doi.org/10.1007/978-3-030-62942-7_5

Other specific function of the AWR is interaction with ground-based responder beacons. New functions of the AWR are detection and visualization of runways at approach landing as well as visualization of taxiways and obstacles on the taxiway at taxiing.

Certainly, not all of the mentioned functions are implemented in a particular airborne radar system. Nevertheless, the AWR always is a multifunctional system that provides earth surface surveillance and weather observation. Usually, weather radar should at least enable to detect clouds and precipitation, select zones of meteorological danger, and show radar image of surface in the map mode.

AWRs or multimode radars with a weather mode are usually nose mounted. Most AWRs operate in either X- or C-band (Kayton and Fried 1997). The λ^{-4} dependence of weather formations on carrier wavelength λ favors X-band radar for their detecting. At the same time, the X-band provides the performance of the long-range weather mode better than the Ku-band. The AWR antenna, in the ground-mapping mode, has a large cosecant-squared elevation beam where the horizontal dimension is narrow ($2° - 6°$) while the other is relatively broad ($10° - 30°$), and it sweeps in an azimuth sector (up to $\pm$ 100°) (Kayton and Fried 1997; Sosnovskiy and Khaymovich 1987; Yanovsky 2003). The scan plane is horizontal because the antenna is stabilized (roll-and-pitch-stabilized). Those features allow enhancing the AWR functionality and use it in the ground-mapping mode as a scatterometer for measuring the water-surface backscattering signature and wind vector over the water surface.

5.2 Measuring Water-Surface Backscattering Signature with Airborne Weather Radar

Let an aircraft equipped with an AWR make a horizontal rectilinear flight with the speed V at some altitude H above the mean sea surface, the AWR operate in the ground-mapping mode as a scatterometer, the radar antenna have different beamwidth in the vertical $\theta_{a.v}$ and horizontal $\theta_{a.h}$ planes ($\theta_{a.v} > \theta_{a.h}$), and scan periodically through an azimuth in a sector as shown in Fig. 5.1. Also, let a delay selection be used to provide a necessary resolution in the vertical plane.

Then, beam scanning allows for selecting power backscattered by the underlying surface for a given incidence angle θ from various directions in an azimuth sector relative to the aircraft course ψ. Angular selection (narrow horizontal beamwidth) in the horizontal plane along with the delay selection provides angular resolutions in the azimuthal and vertical planes, $\Delta\alpha_b$ and $\Delta\theta$, respectively. A current NRCS value obtained from the selected cell is $\sigma°(\theta, \psi + \psi_b)$, where ψ_b is the current azimuth direction of the beam relative to the aircraft current course (right position is positive).

If beam azimuth direction relative to the aircraft current course is fixed, azimuth NRCS curve can be obtained using circle track flight. The fixed beam should be pointed to the outer side of the aircraft turn to observe a greater area of the water surface and to obtain a greater number of independent NRCS samples. From that

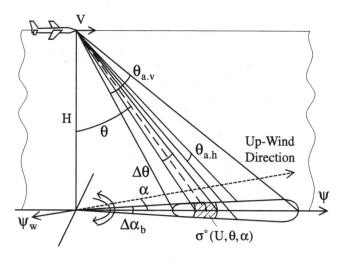

Fig. 5.1 Airborne weather radar beam and selected cell geometry

point of view, the best beam position is when the azimuth direction of the beam is perpendicular to the aircraft current course. If this perpendicular beam position is unavailable due to the AWR narrow scanning sector, the fixed beam should take an outermost position. Depending on the clockwise or anticlockwise aircraft turn the fixed beam should be pointed to the left or right side, respectively. As the scan plane is horizontal because the antenna is stabilized, the aircraft roll should not exceed the maximum allowed for ensuring antenna stabilization and consequently the incidence angle invariability.

Let a horizontal circular flight with the speed V and the right or left roll γ_{fa} at some altitude H above the mean sea surface be completed (Fig. 5.2). Then, the radius of the aircraft turn and the ground range are given by (3.3) and (3.8). The radius of turn of the selected cell middle point is described by (Nekrasov 2011)

$$R_{t.c} = \sqrt{R_{t.fa}^2 + R_g^2 + 2R_{t.fa}R_g \sin \psi_b}. \tag{5.1}$$

The time of the aircraft turn for 360° is given by (3.5).

Let the middle azimuth of the observing sector be the azimuth of the sector, the azimuth size of the sector relative to the center point of circle of the aircraft track be $\Delta\alpha_s$, and the middle azimuth of the sector be α_s. Then, NRCS samples obtained from a sector and averaged over all measurement values in that sector give NRCS value $\sigma^\circ(\theta, \psi_s)$ corresponding to the real observation azimuth angle of the sector ψ_s that is given by (3.6). Real observation azimuth angles of the sector beginning and the sector ending are given by (3.7) and (3.8) at a right aircraft turn and given by (4.11) and (4.12) at a left aircraft turn.

The time of a sector view and the number of samples that can be obtained from a sector are given by (3.9) and (3.10).

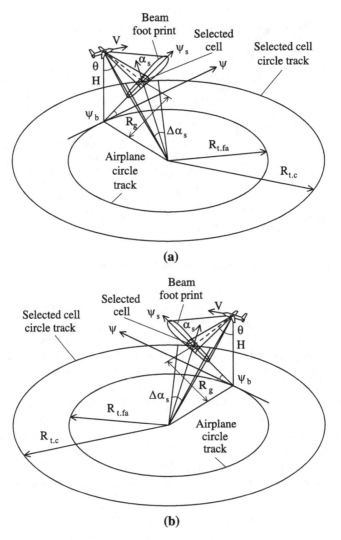

Fig. 5.2 Circular flight geometries for measuring water-surface backscattering signature by the AWR. **a** Clockwise flight. **b** Anticlockwise flight

The measured NRCS data set is in fact discrete, and so each NRCS value obtained is $\sigma°(\theta, \psi_{s.i})$, where $i = \overrightarrow{1, N}$, N is the number of sectors observed during the aircraft turn for $360°$ at the NRCS measurement, $N = 360°/\Delta\alpha_s$.

Thus, to obtain azimuth NRCS curve of the water surface in the range of moderate to high incidence angles under aircraft circular flight by an AWR operating in the ground-mapping mode as a scatterometer, the measurement should be performed in accordance with a scheme shown in Fig. 3.3 at a right aircraft turn, or in accordance to a scheme shown in Fig. 4.5 at a left aircraft turn.

The measurement is started when a stable horizontal circular flight at the given altitude, speed of flight, roll and pitch has been established. The measurement is finished when the azimuth of the measurement starting point is reached. To obtain a greater number of NRCS samples for each sector observed, several consecutive full circle 360° turns should be completed. As the scan plane is horizontal because the antenna is stabilized, the aircraft roll should not exceed the maximum allowed for ensuring antenna stabilization and consequently the incidence angle constancy.

5.3 Wind Retrieval from Azimuth NRCS Data Obtained by Airborne Weather Radar

Let sea-surface backscattering be of the form (2.1). The NRCS model function for medium incidence angles (2.1) can be used without any correction at wind measurement while the azimuth angular size of a cell is up to $15° - 20°$ that is evident from (3.18)–(3.20), Fig. 3.4, and Table 3.2. As measured NRCS data set is also a function of the wind speed, each NRCS value obtained $\sigma°(\theta, \psi_{s.i})$ is considered now as $\sigma°(U, \theta, \psi_{s.i})$ (Nekrasov and Veremyev 2016).

Let the angle between the up-wind direction and the first NRCS azimuth $\psi_{s.1}$ be α, the sector width be $\Delta\alpha_s$, and so the measured NRCSs $\sigma°(U, \theta, \psi_{s.1})$, $\sigma°(U, \theta, \psi_{s.2}),..., \sigma°(U, \theta, \psi_{s.N})$ be the same as $\sigma°(U, \theta, \alpha)$, $\sigma°(U, \theta, \alpha + \Delta\alpha),...,$ $\sigma°(U, \theta, \alpha + (N - 1)\Delta\alpha)$, respectively, at a clockwise measurement, or be the same as $\sigma°(U, \theta, \alpha)$, $\sigma°(U, \theta, \alpha - \Delta\alpha),..., \sigma°(U, \theta, \alpha - (N - 1)\Delta\alpha)$, respectively, at an anticlockwise measurement, where $i = \overline{1, N}$, N is the number of sectors composing the 360° azimuth NRCS curve, $N = 360°/\Delta\alpha_s$. Then, in a general case, to find the wind speed and up-wind direction the system of N Eq. (3.21) corresponding to clockwise measurement case, or the system of N Eq. (4.13) corresponding to anticlockwise measurement case should be solved approximately using searching procedure within ranges of discrete values of possible solutions. Then, navigation wind direction can be found from (3.22). To simplify and significantly speed up the wind vector recovery with (3.21) or (4.13), the procedures previously described in Sect. 4.3 can be used.

It should be noted that in the wind measurement mode an AWR operated in the ground-mapping mode as a scatterometer should use horizontal transmit and receive polarization as the difference in the up-wind and down-wind NRCS values at that polarization is greater than at vertical transmit and receive polarization (Masuko et al. 1986; Moore and Fung 1979; Ulaby et al. 1982). The AWR also should provide the incidence angle of selected cells $\theta \rightarrow 45°$ which is explained by better usage of the anisotropic properties of water-surface scattering at medium incidence angles (Ulaby et al. 1982) as well as by power reasons. For water surface, NRCS falls radically as incidence angle increases and assumes different values for various conditions of sea state or water roughness while, for most other types of terrain, NRCS decreases slowly with the increase of the beam incidence angle (Kayton and Fried 1997). Otherwise, the incidence angle of selected cells should be in the range of validity of the NRCS model function (2.1), and should be out of the "shadow" region of water backscatter.

5.4 Measuring Wind Vector with Airborne Weather Radar at a Rectilinear Flight

The AWR beam scanning through an azimuth sector in the ground-mapping mode allows selecting a power backscattered by the underlying surface from different directions within a sector. At a medium incidence angle, at least three or four NRCS values obtained from considerably different azimuth directions may be quite enough to measure the wind vector over water by the scatterometer method (Nekrassov 1997). Thus, it is desirable that the azimuth scanning sector of the AWR should be as wide as possible for measuring the wind vector over water by the intensity of the reflected signal. Unfortunately, not all AWRs have the azimuth scanning sector of $\pm 90°$ or wider (up to $\pm 100°$), and consequently, this feature should be taken into account when developing measurement algorithms and performing wind measurements.

Depending on the AWR scanning features, three main cases may take place: a narrow scanning sector case (not narrower than $\pm 45°$), a medium scanning sector case (narrower than $\pm 90°$ but wider than $\pm 45°$), and a wide scanning sector case ($\pm 90°$ or wider).

Let an aircraft equipped with the AWR perform a horizontal rectilinear flight with speed V at some altitude H above the mean sea surface, let the AWR operate in the ground-mapping mode as a scatterometer, the radar antenna have different beamwidth in the vertical $\theta_{a.v}$ and horizontal $\theta_{a.h}$ planes ($\theta_{a.v} > \theta_{a.h}$) as shown in Fig. 5.1, and scan periodically through an azimuth in a sector.

Let the sea surface wind blow in direction ψ_w, and the angle between the up-wind direction and the aircraft course ψ is α. Let the NRCS model function for medium incidence angles be of the form (2.1). When the selected cell is narrow enough in the vertical plane, the NRCS model function for medium incidence angles (2.1) can be used without any correction for wind measurement for the azimuth angular size of a cell up to $15°$–$20°$ as mentioned above.

If the scanning sector is narrow but not narrower than $\pm 45°$, the NRCS values may be obtained from three directions $\alpha - 45°$, α, and $\alpha + 45°$ as shown in Fig. 5.3. The NRCS values are $\sigma°(U, \theta, \alpha - 45°)$, $\sigma°(U, \theta, \alpha)$, and $\sigma°(U, \theta, \alpha + 45°)$, respectively. Then, the following algorithm to estimate the wind vector over the sea surface can be proposed (Nekrasov and Labun 2008).

Using measurement geometry and Eq. (2.1), the following system of three equations can be written down

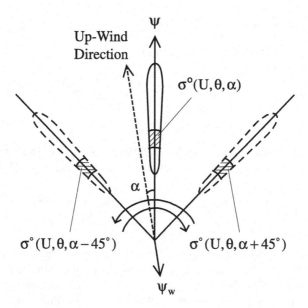

Fig. 5.3 Scanning beam footprints and selected cells in the narrow sector case

$$\begin{cases} \sigma°(U, \theta, \alpha - 45°) = A(U, \theta) \\ \qquad\qquad + B(U, \theta) \cos(\alpha - 45°) \\ \qquad\qquad + C(U, \theta) \cos(2(\alpha - 45°)), \\ \sigma°(U, \theta, \alpha) = A(U, \theta) + B(U, \theta) \cos \alpha \\ \qquad\qquad + C(U, \theta) \cos(2\alpha), \\ \sigma°(U, \theta, \alpha + 45°) = A(U, \theta) \\ \qquad\qquad + B(U, \theta) \cos(\alpha + 45°) \\ \qquad\qquad + C(U, \theta) \cos(2(\alpha + 45°)), \end{cases} \qquad (5.2)$$

and solved approximately using searching procedure within the ranges of discrete values of possible solutions, or analytically. The analytical solution proposed is as follows.

From the sum of the first and the third equations of (5.2), we have

$$\cos \alpha = \frac{\sigma°(U, \theta, \alpha - 45°) + \sigma°(U, \theta, \alpha + 45°) - 2A(U, \theta)}{\sqrt{2}B(U, \theta)}. \qquad (5.3)$$

Using (5.3) and the expression $\cos(2x) = 2\cos^2 x - 1$, the second equation of (5.2) can be presented in the following form

$$\sigma^\circ(U,\theta,\alpha) = (1-\sqrt{2})A(U,\theta)$$
$$+ \frac{1}{\sqrt{2}}(\sigma^\circ(U,\theta,\alpha-45^\circ)+\sigma^\circ(U,\theta,\alpha+45^\circ))$$
$$+ C(U,\theta)\left[\left(\frac{\sigma^\circ(U,\theta,\alpha-45^\circ)+\sigma^\circ(U,\theta,\alpha+45^\circ)-2A(U,\theta)}{B(U,\theta)}\right)^2 - 1\right].$$

$$(5.4)$$

The wind speed over water can be calculated from (5.4). Then, two possible up-wind directions relative to the aircraft course can be found from (5.3). They are

$$\alpha_{1,2} = \pm\arccos\left(\frac{\sigma^\circ(U,\theta,\alpha-45^\circ)+\sigma^\circ(U,\theta,\alpha+45^\circ)-2A(U,\theta)}{\sqrt{2}B(U,\theta)}\right). \qquad (5.5)$$

The unique up-wind direction α relative to the aircraft course can be found by substitution of the values α_1 and α_2 into the first and the third equations of the system of Eq. (5.2). Finally, the wind direction can be found from (4.20).

Another case takes place when the AWR has the scanning sector of medium width that is narrower than $\pm 90^\circ$ but wider than $\pm 45^\circ$ allowing to obtain the NRCS values from five significantly different directions, e.g., from $\alpha-60^\circ$, $\alpha-30^\circ$, α, $\alpha+30^\circ$, and $\alpha+60^\circ$, as shown in Fig. 5.4.

Those NRCSs are $\sigma^\circ(U,\theta,\alpha-60^\circ)$, $\sigma^\circ(U,\theta,\alpha-30^\circ)$, $\sigma^\circ(U,\theta,\alpha)$, $\sigma^\circ(U,\theta,\alpha+30^\circ)$, $\sigma^\circ(U,\theta,\alpha+60^\circ)$, respectively, and another system of equations could be written down (Nekrasov 2013)

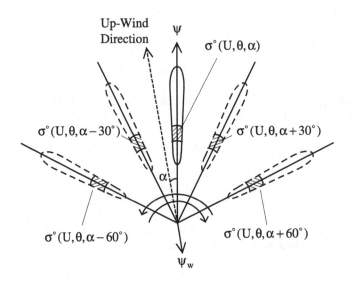

Fig. 5.4 Scanning beam footprints and selected cells in medium sector case

$$
\begin{cases}
\sigma^\circ(U, \theta, \alpha - 60^\circ) = A(U, \theta) \\
\qquad + B(U, \theta)\cos(\alpha - 60^\circ) \\
\qquad + C(U, \theta)\cos(2(\alpha - 60^\circ)), \\
\sigma^\circ(U, \theta, \alpha - 30^\circ) = A(U, \theta) \\
\qquad + B(U, \theta)\cos(\alpha - 30^\circ) \\
\qquad + C(U, \theta)\cos(2(\alpha - 30^\circ)), \\
\sigma^\circ(U, \theta, \alpha) = A(U, \theta) + B(U, \theta)\cos\alpha \\
\qquad + C(U, \theta)\cos(2\alpha), \\
\sigma^\circ(U, \theta, \alpha - 30^\circ) = A(U, \theta) \\
\qquad + B(U, \theta)\cos(\alpha + 30^\circ) \\
\qquad + C(U, \theta)\cos(2(\alpha + 30^\circ)), \\
\sigma^\circ(U, \theta, \alpha - 60^\circ) = A(U, \theta) \\
\qquad + B(U, \theta)\cos(\alpha + 60^\circ) \\
\qquad + C(U, \theta)\cos(2(\alpha + 60^\circ)).
\end{cases}
\tag{5.6}
$$

The system of Eq. (5.6) can be solved approximately using searching procedure within the ranges of discrete values of possible solutions. Then, wind direction can be found from (4.20).

Also, the following analytical solution speeding up calculations for the system of Eq. (5.6) could be written.

From the sum of the first, second, fourth, and fifth equations of (5.6), we have

$$
\cos\alpha = \frac{\begin{array}{l}\sigma^\circ(U, \theta, \alpha - 60^\circ) + \sigma^\circ(U, \theta, \alpha + 30^\circ) \\ \to\ + \sigma^\circ(U, \theta, \alpha + 30^\circ) + \sigma^\circ(U, \theta, \alpha + 60^\circ) - 4A(U, \theta)\end{array}}{(1 + \sqrt{3})B(U, \theta)}.
\tag{5.7}
$$

Using (5.7) and the expression $\cos(2x) = 2\cos^2 x - 1$, the third equation of (5.6) can be presented in the following form

$$
\begin{aligned}
\sigma^\circ(U, \theta, \alpha) = {}& \left(1 - \frac{4}{1 + \sqrt{3}}\right)A(U, \theta) \\
& + \frac{1}{1 + \sqrt{3}}\left(\sigma^\circ(U, \theta, \alpha - 60^\circ) + \sigma^\circ(U, \theta, \alpha - 30^\circ)\right. \\
& \left. + \sigma^\circ(U, \theta, \alpha + 30^\circ) + \sigma^\circ(U, \theta, \alpha + 60^\circ)\right) \\
& + C(U, \theta)\left[2\left(\frac{\begin{array}{l}\sigma^\circ(U, \theta, \alpha - 60^\circ) + \sigma^\circ(U, \theta, \alpha - 30^\circ) \\ \to\ + \sigma^\circ(U, \theta, \alpha + 30^\circ) + \sigma^\circ(U, \theta, \alpha + 60^\circ) - 4A(U, \theta)\end{array}}{(1 + \sqrt{3})B(U, \theta)}\right)^2 - 1\right].
\end{aligned}
\tag{5.8}
$$

The wind speed over water can be calculated from (5.8). Then, two possible up-wind directions relative to the course of the aircraft can be found from (5.7). They are

$$\alpha_{1,2} = \pm \arccos \left(\frac{\begin{array}{l} \sigma^{\circ}(U,\theta,\alpha-60^{\circ}) + \sigma^{\circ}(U,\theta,\alpha+30^{\circ}) \\ \rightarrow +\sigma^{\circ}(U,\theta,\alpha+30^{\circ}) + \sigma^{\circ}(U,\theta,\alpha+60^{\circ}) - 4A(U,\theta) \end{array}}{(1+\sqrt{3})B(U,\theta)} \right).$$

(5.9)

The unique up-wind direction α relative to the aircraft course can be found by substitution of the values α_1 and α_2 into the first and the fifth equations of the system of Eq. (5.6). Then, wind direction can be found from (4.20).

One more case takes place when the AWR beam scans periodically through an azimuth in a wide sector of $\pm 90^{\circ}$ or wider as shown in Fig. 5.5. In this case, the

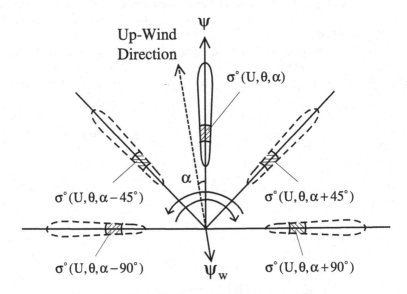

Fig. 5.5 Scanning beam footprints and selected cells in the wide sector case

NRCS values could be obtained from five directions, namely from $\alpha - 90°$, $\alpha - 45°$, α, $\alpha + 45°$, and $\alpha + 90°$. They are $\sigma°(U, \theta, \alpha - 90°)$, $\sigma°(U, \theta, \alpha - 45°)$, $\sigma°(U, \theta, \alpha)$, $\sigma°(U, \theta, \alpha + 45°)$, and $\sigma°(U, \theta, \alpha + 90°)$, respectively. Then, the following algorithm to estimate the wind vector over water can be proposed (Nekrasov 2009).

The following system of equations could be written and solved

$$
\begin{cases}
\sigma°(U, \theta, \alpha - 90°) = A(U, \theta) \\
\qquad + B(U, \theta)\cos(\alpha - 90°) \\
\qquad + C(U, \theta)\cos(2(\alpha - 90°)), \\
\sigma°(U, \theta, \alpha - 45°) = A(U, \theta) \\
\qquad + B(U, \theta)\cos(\alpha - 45°) \\
\qquad + C(U, \theta)\cos(2(\alpha - 45°)), \\
\sigma°(U, \theta, \alpha) = A(U, \theta) + B(U, \theta)\cos\alpha \\
\qquad + C(U, \theta)\cos(2\alpha), \\
\sigma°(U, \theta, \alpha + 45°) = A(U, \theta) \\
\qquad + B(U, \theta)\cos(\alpha + 45°) \\
\qquad + C(U, \theta)\cos(2(\alpha + 45°)), \\
\sigma°(U, \theta, \alpha + 90°) = A(U, \theta) \\
\qquad + B(U, \theta)\cos(\alpha + 90°) \\
\qquad + C(U, \theta)\cos(2(\alpha + 90°)).
\end{cases}
\tag{5.10}
$$

The system of Eq. (5.10) can be solved approximately using a searching procedure within the ranges of discrete values of possible solutions, or analytically. The analytical solution proposed is as follows.

From the sum of the first and the fifth equations of (5.10), we have

$$
\cos 2\alpha = \frac{\sigma°(U, \theta, \alpha - 90°) + \sigma°(U, \theta, \alpha + 90°) - 2A(U, \theta)}{2C(U, \theta)}.
\tag{5.11}
$$

From the sum of the second and the fourth equations of (5.10), we obtain (5.3). Substitution of cos2α from (5.11) and cosα from (5.3) into the third equation of system (5.10) gives the following formula

$$A(U, \theta) = -\frac{1}{\sqrt{2}} \sigma°(U, \theta, \alpha)$$

$$+ \frac{1}{2}(\sigma°(U, \theta, \alpha - 45°) + \sigma°(U, \theta, \alpha + 45°)) \qquad (5.12)$$

$$+ \frac{1}{2\sqrt{2}}(\sigma°(U, \theta, \alpha - 90°) + \sigma°(U, \theta, \alpha + 90°)).$$

The wind speed over water can be calculated from (5.12). Then, two possible up-wind directions relative to the course of the flying apparatus α_1 and α_2 can be found from (5.5). The unique up-wind direction α relative to the course can be found by substitution of the values α_1 and α_2 into the first and the fifth equations of the system of Eq. (5.10). Finally, the wind direction can be found from (4.20).

The analytical solutions proposed can speed up calculations and their results could also be used for more precise calculations using the systems of Eqs. (5.2), (5.6), and (5.10) depending on the appropriate AWR features.

Wind measurement is started when a stable rectilinear flight at the given altitude and speed of flight has been established. The measurement is finished when a required number of NRCS samples for each significantly different azimuth direction is obtained. To obtain a greater number of NRCS samples for each direction observed, several consecutive beam sweeps should be used.

Full advantage of the wide scanning sector ($\pm 90°$ or wider, up to $\pm 100°$) AWR geometry at a rectilinear flight can be achieved when the wide scanning sector is divided for many narrow azimuth sectors with the sector azimuth width $\Delta\alpha_s$. In this case, NRCS samples are obtained from each narrow sector and averaged over all measured values in the sector give an appropriate integrated NRCS value corresponding to the azimuth angle of the narrow sector. A number of the narrow sectors formed in the wide scanning sector is $N = 180°/\Delta\alpha_s + 1$. Thus, N NRCSs can be obtained from significantly different azimuth angles, and a system of N equations of form (2.1) can be written down.

Let the width of a narrow sector be 5°. Then, the number of narrow sectors is 37, and the system of 37 equations can be written down (Nekrasov et al. 2016).

$$
\left\{
\begin{aligned}
\sigma^\circ(U, \theta, \alpha - 90^\circ) &= A(U, \theta) \\
&\quad + B(U, \theta) \cos(\alpha - 90^\circ) \\
&\quad + C(U, \theta) \cos(2(\alpha - 90^\circ)), \\
\sigma^\circ(U, \theta, \alpha - 85^\circ) &= A(U, \theta) \\
&\quad + B(U, \theta) \cos(\alpha - 85^\circ) \\
&\quad + C(U, \theta) \cos(2(\alpha - 85^\circ)), \\
&\cdots \\
\sigma^\circ(U, \theta, \alpha - 45^\circ) &= A(U, \theta) \\
&\quad + B(U, \theta) \cos(\alpha - 45^\circ) \\
&\quad + C(U, \theta) \cos(2(\alpha - 45^\circ)), \\
&\cdots \\
\sigma^\circ(U, \theta, \alpha) &= A(U, \theta) \\
&\quad + B(U, \theta) \cos \alpha \\
&\quad + C(U, \theta) \cos(2\alpha), \\
&\cdots \\
\sigma^\circ(U, \theta, \alpha + 45^\circ) &= A(U, \theta) \\
&\quad + B(U, \theta) \cos(\alpha + 45^\circ) \\
&\quad + C(U, \theta) \cos(2(\alpha + 45^\circ)), \\
&\cdots \\
\sigma^\circ(U, \theta, \alpha + 85^\circ) &= A(U, \theta) \\
&\quad + B(U, \theta) \cos(\alpha + 85^\circ) \\
&\quad + C(U, \theta) \cos(2(\alpha + 85^\circ)), \\
\sigma^\circ(U, \theta, \alpha + 90^\circ) &= A(U, \theta) \\
&\quad + B(U, \theta) \cos(\alpha + 90^\circ) \\
&\quad + C(U, \theta) \cos(2(\alpha + 90^\circ)),
\end{aligned}
\right.
\tag{5.13}
$$

where $\sigma^\circ(U, \theta, \alpha - 90^\circ), \ldots, \sigma^\circ(U, \theta, \alpha), \ldots,$ and $\sigma^\circ(U, \theta, \alpha + 90^\circ)$ are the NRCSs obtained for appropriate narrow sectors corresponded to directions $\alpha - 90^\circ, \ldots, \alpha, \ldots,$ and $\alpha + 90^\circ$, respectively.

The system of Eq. (5.13) can be solved approximately using the searching procedure within the ranges of discrete values of possible solutions. And then, the navigation wind direction can be found from Eq. (4.20).

Thus, a fast mode and a normal mode could be realized for AWR with the wide scanning sector. The normal mode is based on the NRCS data obtained for all the directions $\alpha - 90^\circ, \ldots, \alpha, \ldots,$ and $\alpha + 90^\circ$ observed within a wide scanning sector, and the wind retrieval performs with the system of Eq. (5.13). The fast mode is based on 5 azimuth NRCSs measurement obtained for azimuthal directions $\alpha - 90°, \alpha - 45°, \alpha, \alpha + 45°, \alpha + 90°$ by solving the system of Eq. (5.10) with the help of Eq. (5.12).

The wind speed and direction retrieval also can be speeded up by narrowing the range of possible wind speeds. A lower wind speed U_L and an upper wind speed U_U can be found using an averaged 180° azimuth NRSCs $\sigma°(U, \theta, \alpha, \Delta\alpha_w = 180°)$ from the following equations (Nekrasov et al. 2016).

$$\sigma°(U, \theta, \alpha, \Delta\alpha_w = 180°) = A(U_L, \theta) + \frac{2}{\pi}B(U_L, \theta), \qquad (5.14)$$

$$\sigma°(U, \theta, \alpha, \Delta\alpha_w = 180°) = A(U_U, \theta) - \frac{2}{\pi}B(U_U, \theta). \qquad (5.15)$$

Thus, the fast mode results can be used when needed as estimates for a more precise calculation of the wind speed and direction in the normal mode.

Assuming the requirement that the wind and wave conditions in different parts of the observed area be identical for all its parts, the measurement swath width as well as the length of the area observed, should be no larger than 15–20 km. This requirement leads to an altitude limitation for the method's applicability by AWR. The maximum altitude limitation at a rectilinear flight will be about 10 km at the incidence angle of 45° and 5 km at the incidence angle of 60°, respectively.

Despite the fact that AWR with the wide scanning sector cannot provide a whole 360° azimuth NRSC measurement at the rectilinear flight for one stage, a two-stage procedure can also be used for that purpose (Nekrasov and Popov 2015; Nekrasov and Dell'Acqua 2016). In that case, the two-stage measurement should be performed in accordance with a scheme presented in Fig. 5.6 to obtain two 180° azimuth NRCS data sets which are used to form the entire 360° azimuth NRCS data set and estimate the wind speed and direction over the sea surface in accordance with the wind retrieval algorithm based on Eq. (3.21)–(3.25).

The first stage of the measurement is started when a stable horizontal rectilinear flight at the given altitude and speed of flight has been established. The first stage is finished when a required number of NRCSs for each observed narrow azimuth sector is obtained. The second stage of the measurement is started when the aircraft has turned for 180° and a stable horizontal rectilinear flight at the given altitude and speed of flight has been established again. The second stage is finished when a required number of NRCSs for each narrow azimuth sector observed is obtained. To obtain a greater number of NRCS samples for each narrow sector, several consecutive two-stage measurements also can be performed.

The joint two-stage measurement swath width, as well as the length of the area observed (distance between the start and end point of the stage), should be no larger than 15–20 km to assume the wind and wave conditions in different parts of the observed area to be identical for all its parts.

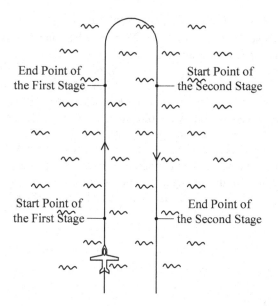

Fig. 5.6 Two-stage scheme for the measurement of the water-surface backscattering signature at the rectilinear flight

5.5   Simulation of Wind Vector Retrieval by Airborne Weather Radar at a Rectilinear Flight

To verify the feasibility of the wind measurement algorithm for AWR operated in the ground-mapping mode as a scatterometer scanning in a wide sector at the rectilinear flight a simulation based on Eq. (5.13) has been performed. As an X-band geophysical model function of the form of Eq. (2.1) for the horizontal transmit and receive polarization is unavailable in current literature, a Ku-band geophysical model function of the form of Eq. (2.1) for the required transmit and receive polarization from (Moore and Fung 1979) has been used for this simulation. Based on (Masuko et al. 1986), such substitution is quite applicable at the verification of the algorithm as only modest differences between the X- and Ku-band backscatters take place.

An incidence angle of 45° has been considered at simulation. The "measured" azimuth NRCSs were generated using a Rayleigh Power (Exponential) distribution. An azimuth NRCS curve corresponding to a "true" wind speed of 10 m/s and up-win direction of 0° (Figs. 5.7, 5.8 and 5.9, solid traces) has been used for generation of the "measured" NRCSs taking into account a 0.2 dB instrumental noise at the simulation. 1565 "measured" NRCS samples have been averaged for each five-degree narrow azimuth sector (Figs. 5.7, 5.8, and 5.9, dot traces). The "measured" wind speeds and up-wind directions have been calculated using the system of Eq. (5.3) and "measured" NRCS values obtained for the following 180° azimuth sectors: [−90°, 90°] (Figs. 5.7, 5.8, and 5.9 dash trace), [45°, 225°] (Figs. 5.7, 5.8, and 5.9 dash trace), and [90°, 270°] (Fig. 5.9 dash trace). The "measured" wind speeds and up-wind directions for these wide sectors were 9.994 m/s and 357.9°, 10.0258 m/s and 356.9°, 10.0279 m/s and 357.3°, respectively.

The simulation results clearly indicate the suitability of the AWR operated in the ground-mapping mode as a scatterometer scanning in a wide sector for such measurements at the rectilinear flight, and the accuracy of the proposed algorithm, even at the worst case of the 180° azimuth sector location of [90°, 270°].

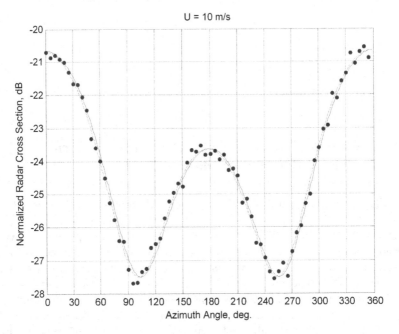

Fig. 5.7 Azimuth NRCS curve using Eq. (2.1) at the incidence angle of 45°, "true" wind speed of 10 m/s and up-wind direction of 0° (solid trace); generated "measured" NRCS with taking into account the instrumental noise of 0.2 dB after averaging of 1565 NRCSs in a five-degree azimuth sector (dot trace); and azimuth NRCS curve using Eq. (2.1) corresponding to "measured" wind speed of 9.994 m/s and up-wind direction of 357.9° retrieved from the azimuth sector of [−90°, 90°] (dash trace)

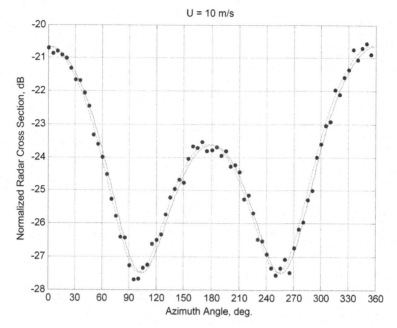

Fig. 5.8 Azimuth NRCS curve using Eq. (2.1) at the incidence angle of 45°, "true" wind speed of 10 m/s and up-wind direction of 0° (solid trace); generated "measured" NRCS with taking into account the instrumental noise of 0.2 dB after averaging of 1565 NRCSs in a five-degree azimuth sector (dot trace); and azimuth NRCS curve using Eq. (2.1) corresponding to "measured" wind speed of 10.0258 m/s and up-wind direction of 356.9° retrieved from the azimuth sector of [45°, 225°] (dash trace)

A similar simulation result is obtained at the circular or two-stage rectilinear flight for the wind retrieval by AWR. An example of such a simulation based on system of Eq. (3.21) at the same "true" wind speed of 10 m/s and up-wind direction of 0° but with a reduced number of integrated NRCS samples from the previous case of 1565 to 315 for each five-degree azimuth sector is presented in Fig. 5.10. The following "measured" wind speed and up-wind direction of 9.9913 m/s and 357.4° have been obtained as the result.

The simulation example also clearly indicate the suitability of the AWR operated in the ground-mapping mode as a scatterometer scanning in a wide sector for such measurements at the circular flight as well as at the two-stage rectilinear flight, and the accuracy of the proposed algorithm.

The superiority of the one-stage rectilinear flight measurement in comparison with the circular and at two-stage rectilinear flights is that it is much easier to a pilot and shorter that especially important at safe and rescue missions compared to operational research measurements.

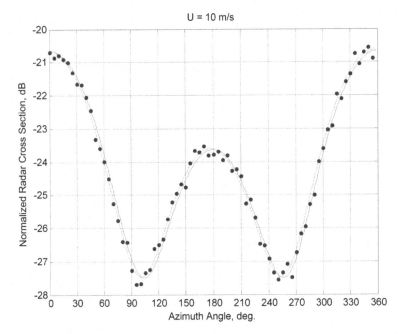

Fig. 5.9 Azimuth NRCS curve using Eq. (2.1) at the incidence angle of 45°, "true" wind speed of 10 m/s and up-wind direction of 0° (solid trace); generated "measured" NRCS with taking into account the instrumental noise of 0.2 dB after averaging of 1565 NRCSs in a five-degree azimuth sector (dot trace); and azimuth NRCS curve using Eq. (2.1) corresponding to "measured" wind speed of 10.0279 m/s and up-wind direction of 357.3° retrieved from the azimuth sector of [90°, 270°] (dash trace)

5.6 Conclusions to Measuring Water-Surface Backscattering Signature and Wind by Means of Airborne Weather Radar

The analysis of AWR has shown that the radar employed in the ground-mapping mode as a scatterometer can be used for remote measuring of the sea-surface backscattering signature at a circular flight along with recovering the wind speed and direction over the water surface from NRCS azimuth curves obtained as well as for measuring wind vector during a rectilinear flight in addition to its typical meteorological and navigation application.

The azimuth NRCS curve is obtained with AWR during a circular track flight when the azimuth direction of the beam relative to the aircraft current course is fixed. The fixed beam should be pointed to the outer side of an aircraft turn to observe a greater area of the water surface and to obtain a greater number of independent NRCS samples. Azimuth beam direction should be perpendicular to the aircraft current course or at least tend to perpendicular position when the scanning sector is narrower than ±90°. As the scan plane is horizontal because the antenna is

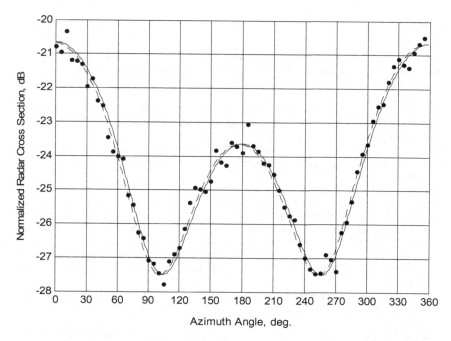

Fig. 5.10 Azimuth NRCS curve using Eq. (2.1) at the incidence angle of 45°, "true" wind speed of 10 m/s and up-wind direction of 0° (solid trace); generated "measured" NRCS with taking into account the instrumental noise of 0.2 dB after averaging of 315 NRCSs in a five-degree azimuth sector (dot trace); and azimuth NRCS curve using Eq. (2.1) corresponding to "measured" wind speed of 9.9913 m/s and up-wind direction of 357.4° (dash trace)

stabilized, the aircraft roll should not exceed the maximum allowed value for ensuring antenna stabilization and consequently incidence angle invariability.

AWR also can be used for measuring wind vector over water during a rectilinear flight. Depending on width of scanning sector, appropriate measuring geometry and algorithm should be used to obtain NRCS data from the widest possible scanning sector for given AWR. The wide scanning sector of ±90° or wider is preferable in comparison with a medium or especially narrow sector, as it allows to obtain NRCS values from significantly different azimuth directions which provides better wind vector estimation.

At the near-surface wind measurement, AWR operated in the scatterometer mode should use the horizontal transmit and receive polarization as the difference in the up-wind and down-wind NRCS values at that polarization is greater than at vertical transmit and receive polarization. It should also provide for incidence angle of selected cells $\theta \rightarrow 45°$ which is explained by better usage of the anisotropic properties of water-surface scattering at medium incidence angles as well as by power reasons. Otherwise, the incidence angle of selected cells should be in the range of validity of the NRCS model function (2.1) and should be out of the "shadow" region of water backscatter.

References

Kayton M, Fried WR (1997) Avionics navigation systems. Wiley, New York, p 773

Masuko H, Okamoto K, Shimada M, Niwa S (1986) Measurement of microwave backscattering signatures of the ocean surface using X band and Ka band airborne scatterometers. J Geophys Res 91(C11):13065–13083

Moore RK, Fung AK (1979) Radar determination of winds at sea. Proc IEEE 67(11):1504–1521

Nekrasov A, Labun J (2008) About measurement of the sea surface wind vector by the airborne weather radar. Acta Avionica X(16):98–103

Nekrasov A (2009) Measurement of the sea surface wind vector by the airborne weather radar having a wide scanning sector. Proceedings of RADAR 2009, Bordeaux, France, 12–16 Oct 2009, p 4

Nekrasov A (2011) Airborne weather radar application for measurement of the water surface backscattering signature. Proceedings of RADAR 2011, Chengdu, China, 24–27 Oct 2011, p 4

Nekrasov A (2013) Water-surface wind vector estimation by an airborne weather radar having a medium-size scanning sector. Proceedings of the 14th International Radar Symposium IRS 2013, Dresden, Germany, 19–21 Jun 2013, pp 1079–1084

Nekrassov A (1997) Measurement of sea surface wind speed and its navigational direction from flying apparatus. In: Proceedings of Oceans'97, Halifax, Nova Scotia, Canada, 6–9 Oct 1997, pp 83–86

Nekrasov A, Dell'Acqua F (2016) Airborne weather radar: theoretical approach for water-surface backscattering and wind measurements with airborne weather radar. IEEE Geosc Rem Sen M 4 (4):38–50

Nekrasov A, Popov D (2015) A concept for measuring the water-surface backscattering signature by airborne weather radar. In: Proceedings of the 16th international radar symposium IRS 2015, Dresden, Germany, 24–26 June 2015, vol 2, pp 1112–1116.

Nekrasov A, Veremyev V (2016) Airborne weather radar concept for measuring water surface backscattering signature and sea wind at circular flight. Nase More 63(4):278–282

Nekrasov A, Khachaturian A, Veremyev V, Bogachev M (2016) Sea surface wind measurement by airborne weather radar scanning in a wide-size sector. Atmosphere 7(5), 72:1–11

Sosnovskiy AA, Khaymovich IA (1987) Radio-electronic equipment of flying apparatuses. Transport, Moscow, USSR, p 256 (in Russian)

Sosnovsky AA, Khaymovich IA, Lutin EA, Maximov IB (1990) Aviation radio navigation: Handbook. Transport, Moscow, USSR, p 264 (in Russian)

Ulaby FT, Moore RK, Fung AK (1982) Microwave remote sensing: active and pasive, volume 2: radar remote sensing and surface scattering and emission theory. Addison-Wesley, London, p 1064

Yanovsky FJ (2003) Meteorological and navigation radar systems of aircrafts, National Aviation University, Kiev, p 304 (in Ukrainian)

Yanovsky FJ (2005) Evolution and prospects of airborne weather radar functionality and technology. Proceedings of ICECom 2005, Dubrovnik, Croatia, 11–14 Oct 2005, pp 349–352

Chapter 6
Water-Surface Wind Retrieval Using Airborne Radar Altimeter

6.1 Airborne Radar Altimeter

Radar altimeters are frequently used by aircraft. The primary function of the ARA is to provide terrain clearance or altitude with respect to the ground level directly beneath the airplane or helicopter. The ARA may also provide a vertical rate of climb or descent and selectable low altitude warning (Kayton and Fried 1997). Thus, the ARA is an essential part in ground proximity warning systems, warning the pilot if the aircraft is flying too low or descending too quickly, especially in low-visibility conditions and also for automatic landings, allowing the autopilot to know when to begin the flare maneuver. As ARAs cannot see terrain directly ahead of the aircraft but only directly below it, such functionality requires either knowledge of the position and the terrain at that position or a forward-looking terrain radar which uses technology similar to a radio altimeter.

Altimeters perform a basic function of any range measuring radar. A modulated signal is transmitted toward the ground. Modulation provides a time reference to which the reflected return signal can be reflected, thereby providing radar-range or time delay and therefore altitude. The ground presents an extended target, as opposed to a point target, resulting in a delay path extending from a point directly beneath the aircraft out to the edge of antenna beam. Furthermore, the beam width of a dedicated radar altimeter antenna must be wide enough to accommodate normal roll-and-pitch angles of the aircraft, resulting in a significant variation in return delay.

The ARA is constructed as FM-CW or pulsed radar. The frequency band of 4.2–4.4 GHz is assigned to the ARA. The frequency is high enough to result in reasonably small sized antennas to produce a $40° - 50°$ beam but is sufficiently low so that rain attenuation and backscatter from rain have no significant range limiting effects. Typical installations include a pair of small microstrip antennas for transmit and receive functions (Kayton and Fried 1997).

© The Author(s), under exclusive license to Springer Nature Switzerland AG 2021
A. Nekrasov, *Foundations for Innovative Application of Airborne Radars*,
SpringerBriefs in Earth Sciences, https://doi.org/10.1007/978-3-030-62942-7_6

6.2 Beam Sharpening

Typical scatterometer wind measurements are commonly performed using antennas with comparatively narrow beams (beamwidth of $4°-10°$). As the ARA has a wide-beam antenna, the beam sharpening technologies should be used to apply the ARA for wind vector measurement.

To sharpen the effective antenna beams of real-aperture radars avoiding the size enlargement of their antennas, Doppler discrimination along with range discrimination have been employed. An example of an application for such a simultaneous range Doppler discrimination technique is the conically scanning pencil-beam scatterometer performing wind retrieval (Spencer et al. 2000a). When simultaneous range Doppler processing is used, the resolution cell is delineated by the iso-Doppler and iso-range lines projected on the surface, where the spacing between the lines is the achievable Doppler or range resolution, respectively. As the beam scans, azimuth resolution is best at the side-looking locations and is coarsest at the forward and afterward locations. A conceptual description of such a scatterometer has been described in (Spencer et al. 2000b).

Another example of employing simultaneous range Doppler discrimination technique is the delay Doppler-radar altimeter developed at the Applied Physics Laboratory of the Johns Hopkins University (Raney 1998). The delay Doppler altimeter uses coherent processing over a block of received returns to estimate the Doppler frequency modulation imposed on the signals by the forward motion of the altimeter. Doppler analysis of the data allows estimating their along-track positions relative to the position of the altimeter. It follows that the along-track dimension of the signal data and the cross-track (range or time delay) dimensions are separable. In contrast to the response of a conventional altimeter having only one independent variable (time delay), the delay Doppler altimeter response has two independent variables: along-track position (functionally related to Doppler frequency) and cross-track position (functionally related to time delay). After delay Doppler processing, these two variables describe an orthonormal data grid. With this data space in mind, delay Doppler processing may be interpreted as an operation that flattens the radiating field in the along-track direction. Unfortunately, a cross-track ambiguity takes place under measurements, as there are two possible sources of reflections (one from the left side and another from the right side), which have a given time delay at any given Doppler frequency (Raney 1998).

Also, the sensitivity of signals from the GPS to propagation effects was found to be useful for measurements of surface roughness characteristics from which wave height, wind speed, and direction could be determined. The Delay Mapping Receiver (DMR) was designed, and a number of airborne experiments were completed. The DMR includes two low-gain (wide-beam) L-band antennas: a zenith mounted right-hand circular polarized antenna, and a nadir mounted left-hand circular polarized (LHCP) antenna. It is assumed that a downward-looking LHCP antenna intercepts only the scattered signal and is insensitive to the direct signal. By combining code-range and Doppler measurements, the receiver distinguished particular patches of the ocean surface illuminated by GPS signal that, in fact, is the delay Doppler spatial selection. The estimated wind speed using surface-reflected GPS data collected at a variety of wind speed conditions showed an overall agreement better than 2 m/s with data obtained from nearby buoy data and independent wind speed measurements derived from satellite observations. Wind direction agreement with QuikSCAT measurements appeared to be at the 30° level (Komjathy et al. 2000, 2001).

The latest example of the delay Doppler technique implementation is a Cryosat-2 delay Doppler altimeter. Further delay Doppler altimeter will be used during Sanitel-3 mission devoted to the provision of operational oceanographic services within the European Earth monitoring program of Global Monitoring for the Environment and Security (Martin-Puig and Ruffini 2009; Gommenginger et al. 2012).

6.3 Wind Vector Estimation Using an Airborne Radar Altimeter with Antenna Forming the Circle Footprint

To obtain operational information on sea wave height and wind vector over water, a radar altimeter and a scatterometer are both required on board of an aircraft. Therefore, their measuring ability can be combined in one instrument. One of the promising ways of such a combination is using a short-pulse wide-beam nadir-looking radar like an airborne Wind-Wave Radar (Hammond et al. 1977) but with two additional Doppler filters. Here, only a short-pulse scatterometer mode for estimating wind vector by such an airborne altimeter is considered.

Let an aircraft equipped with a scatterometer (altimeter) having a nadir-looking wide-beam antenna perform a horizontal rectilinear flight with the speed V at some altitude H above the mean sea surface, the antenna have the same beamwidth θ_a in both the vertical and horizontal planes, forming a glistening zone on the sea surface, and then transmit a short pulse of duration τ at some time $t = 0$ (Fig. 6.1). If the surface is (quasi-) flat, the first signal return, from the nadir point, occurs at time $t_0 = 2H/c$, where c is the speed of light. The trailing edge of the pulse undergoes the same interactions as the leading edge but delayed in time by τ. The last energy is received from nadir at time $t_0 + \tau$, and the angle for the pulse-limited footprint is usually assumed to be as $\theta_p = \sqrt{c\tau/H}$ (Hammond et al. 1977), but exactly it is as follows (Nekrassov 2002)

$$\theta_p = \arccos\left(\frac{H}{H + 0.5\tau c}\right).\tag{6.1}$$

For larger values of time, an annulus is illuminated. The angular incidence resolution $\Delta\theta$ is the poorest at nadir, and it improves rapidly with the time from the nadir point

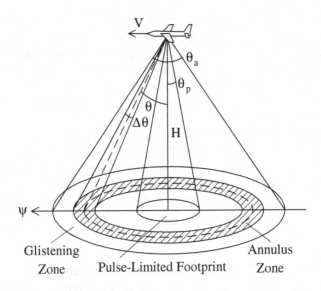

Fig. 6.1 Geometry for measuring the wind speed

$$\Delta\theta = \theta_e - \theta_i, \tag{6.2}$$

where θ_e and θ_i are the incidence angles corresponding to the angular location of the external and internal borders of the annulus (the leading and trailing edges of the pulse)

$$\theta_e = \arccos\left(\frac{H}{H/\cos\theta_i + 0.5\tau c}\right), \tag{6.3}$$

$$\theta_i = \arccos\left(\frac{H}{H + 0.5\Delta tc}\right), \tag{6.4}$$

Δt is the time passed after receiving the last energy from the nadir point.

The incidence angle corresponding to the annulus is determined as the following

$$\theta = 0.5(\theta_e + \theta_i). \tag{6.5}$$

The sea surface can be assumed to be rough, and if waves have some height h_w, the first energy comes back from nadir at time $t_0 = (2H - h_w)/c$, the last energy is received from nadir at time $t_0 + 2h_w/c + \tau$, and the angle for the pulse-limited footprint is

$$\theta_p = \arccos\left(\frac{H - 0.5h_w}{H + 0.5h_w + 0.5\tau c}\right). \tag{6.6}$$

In that case, the angular incidence resolution and the incidence angle corresponding to the annulus are also of the form (6.2) and (6.5) but taking into account that

$$\theta_e = \arccos\left(\frac{H - 0.5h_w}{(H + 0.5h_w)/\cos\theta_i + 0.5\tau c}\right),\tag{6.7}$$

$$\theta_i = \arccos\left(\frac{H + 0.5h_w}{H + 0.5h_w + 0.5\Delta tc}\right).\tag{6.8}$$

From (6.1–6.5), we can see that the pulse duration should be as short as possible (nanoseconds), and the altitude of flight should be at least 50 times higher than the roughness height, which can be taken into account from the wavehight measurement with the altimeter. For example, at $H = 250$ m, $\tau = 10$ ns, $h_w = 5$ m, and $\theta_i = 20°$, we have $\theta_p = 12,2°$, $\theta_e = 24°$, and $\Delta\theta = 4°$ Increasing the altitude of flight leads to a decrease in angle of the pulse-limited footprint and the angular size of the annulus. As typical airborne radar altimeter, which measures the altitude of flight, has a wide-beam (low-gain) antenna, it allows locating the glistening zone quite far from the nadir point for using it for wind measurement.

Let the NRCS model function for medium incidence angles (annulus zone) be of the form (2.1) and the NRCS model function for the pulse-limited footprint be of the form (2.6). Then, the following algorithm to estimate the wind vector over the sea surface can be proposed.

Wind speed can be obtained by means of nadir measurement, for instance, from (2.6) and then converted to a height of measurement of 10 m ($U_{10} = U$), which is mostly used today. For a neutral stability wind profile, the following expression is used for such conversion (Jackson et al. 1992)

$$U_{10} = 0.93U_{19.5}.\tag{6.9}$$

It was obtained by means of the following widely used equation

$$\frac{U_z}{U_{10}} = \left(\frac{z}{10}\right)^{0.13},\tag{6.10}$$

where U_z is the measured wind speed at the anemometer height of z meters.

Using (6.9), the 10-meter wind speed for the NRCS model function (2.6) is

$$U = 0.93 \cdot 10^{[\log_{10}\sigma°(U,0°) - G_1]/G_2}.\tag{6.11}$$

It is necessary to note that the dependence of measured NRCS value on the angular size of a pulse-limited footprint should be taken into account when the narrow-beam NRCS model function is used.

For near-nadir incident angles, where the sea-surface slopes are nearly Gaussian and isotropic in their distribution, the expression for the NRCS is as follows (Hammond et al. 1977)

$$\sigma^\circ(\theta) = \frac{|R(0^\circ)|^2}{\sigma_\gamma^2} \sec^4 \theta \exp\left(-\frac{\tan^2 \theta}{\sigma_\gamma^2}\right),$$ (6.12)

where $R(0^\circ)$ is the Fresnel reflection coefficient at normal incidence, σ_γ^2 is the averaged variance of the sea-surface slopes.

If a wide-beam antenna is used to measure the NRCS that changes rapidly, the narrow-beam approximation error may be considerable. Therefore, it is necessary to reduce the error to an acceptable level when the narrow-beam approximation condition is violated in the measurement (Wang and Gogineni 1991).

Let the NRCS model function for the incident angle of 0° be of the form (2.6). Let the near-nadir backscatter be provided from a pulse-limited footprint that is equivalent to measurement by an antenna with the same beamwidth $2\theta_p$ in both the vertical and horizontal planes.

Then, using (2.6) and (6.12), the expression for the NRCS $\sigma_w^\circ(0^\circ)$ measured by a scatterometer having a nadir-looking wide-beam antenna can be obtained from the following equation (Nekrassov 2001)

$$\sigma_w^\circ(0^\circ) = \frac{1}{2\pi\theta_p} \int_0^{2\pi} \int_0^{\theta_p} \sigma^\circ(\theta) d\theta d\alpha$$

$$= \frac{|R(0^\circ)|^2}{\theta_p \sigma_\gamma^2} \int_0^{\theta_p} \frac{\exp\left(-\frac{\tan^2 \theta}{\sigma_\gamma^2}\right)}{\cos^4 \theta} d\theta = \sigma^\circ(0^\circ) K_w,$$ (6.13)

where K_w is the coefficient taking into account an expansion of the antenna beamwidth,

$$K_w = \frac{1}{2\theta_p}\left[-\tan\theta_p \sigma_\gamma^2 \exp\left(-\frac{\tan^2 \theta_p}{\sigma_\gamma^2}\right) \right.$$

$$\left. + \sqrt{\pi}\sigma_\gamma\left(1 + \frac{\sigma_\gamma^2}{2}\right) erf\left(\frac{\tan\theta_p}{\sigma_\gamma}\right)\right].$$ (6.14)

Analysis of Eq. (6.14) has shown that one can be simplified, and the coefficient taking into account the simplification $K_{w.s}$ can be represented in the following form

$$K_{w.s} = \frac{\sqrt{\pi}\sigma_\gamma}{2\theta_p} erf\left(\frac{\tan\theta_p}{\sigma_\gamma}\right).$$ (6.15)

The dependence of the relative error for the approximation

$$\delta K_{w.s} = \frac{K_{w.s} - K_w}{K_w}$$ (6.16)

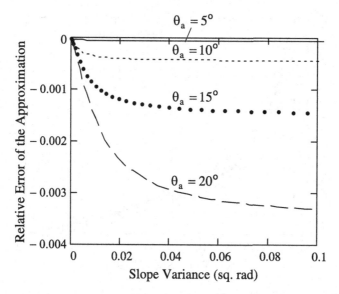

Fig. 6.2 Relative error of the approximation versus averaged variance of the sea-surface slopes for antenna beamwidths $2\theta_p$ of $5°, 10°, 15°,$ and $20°$

on the variance of sea-surface slopes is represented by Fig. 6.2. The figure shows clearly that the relative error of the proposed approximation is quite insignificant.

Thus, measured NRCS can be written as the follows

$$\sigma°_w(0°) = \frac{\sqrt{\pi}|R(0°)|^2}{2\theta_p\sigma_\gamma} erf\left(\frac{\tan\theta_p}{\sigma_\gamma}\right), \qquad (6.17)$$

where $|R(0°)|^2$ is taken equal to 0.631 in (Hammond et al. 1977).

Figure 6.3 demonstrates the effect of antenna beamwidth expansion (increasing pulse duration), which should be taken into account to correct a measurement result. For this reason, the averaged variance of the sea-surface slopes should be calculated from Eq. (6.17) to obtain the NRCS value for the narrow-beam nadir model function with the following expression

$$\sigma°(0°) = \frac{|R(0°)|^2}{\sigma_\gamma^2}. \qquad (6.18)$$

Therefore, nadir NRCS data obtained by an altimeter having a nadir-looking wide-beam antenna should be corrected in case of a pulse-limited footprint angular size being over approximately $5° - 6°$.

Alternatively, wind speed can be obtained from off-nadir measurement at medium incidence angles using (2.1), when average azimuthally integrated NRCS obtained from the annulus zone $\sigma°_{an}(U, \theta)$ is represented as follows

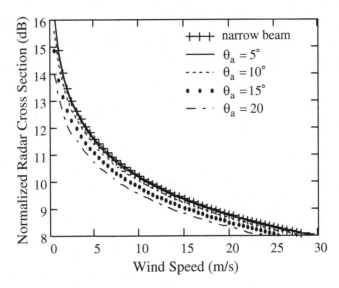

Fig. 6.3 Normalized radar cross section versus wind speed for a narrow-beam antenna and antenna beamwidths $2\theta_p$ of 5°, 10°, 15°, and 20°

$$\sigma^{\circ}_{an}(U, \theta) = \frac{1}{2\pi} \int_{0}^{2\pi} \sigma^{\circ}(U, \theta, \alpha)d\alpha = A(U, \theta), \qquad (6.19)$$

and then, the wind speed can be found from the following equation

$$U = \left(\frac{A(U, \theta)}{a_0(\theta)}\right)^{1/\gamma_0(\theta)} = \left(\frac{\sigma^{\circ}_{an}(U, \theta)}{a_0(\theta)}\right)^{1/\gamma_0(\theta)}. \qquad (6.20)$$

Now assume that narrow enough Doppler zones could be obtained by means of Doppler filtering (Fig. 6.4). Then, the intersection of an annulus with a Doppler zone would form a spatial cell that discriminates the signal scattered back from the appropriate area of the annulus in the azimuthal direction. Employing Doppler filtering, which provides the azimuthal selection under the measurement with the azimuth resolution (azimuth angular size of a cell) $\Delta\alpha$ in the directions of ψ_c, $180° - \psi_c$, $180° + \psi_c$ and $360° - \psi_c$ relative to the aircraft course, where ψ_c is the angle between the aircraft course and the azimuth of the first Doppler cell, the wind direction can be derived. Since the azimuth NRCS curve at medium incidence angle has the principal maximum, the second maximum, and two minima, it is desirable that $\psi_c \rightarrow 45°$ to provide a maximum possible azimuth angle of 90° between the next cells.

To provide for the required azimuth angular size of the cells, the frequency limits for the fore-Doppler filter $F_{D1.f}$ and $F_{D2.f}$ and for the aft-Doppler filter $F_{D1.a}$ and $F_{D2.a}$ (relative to the zero-Doppler frequency shift) should be as follows

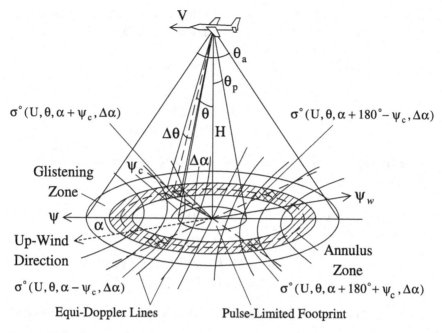

Fig. 6.4 Four-cell geometry for wind vector measurement with an antenna forming a *circle* footprint

$$F_{D1.f} = \frac{2V}{\lambda} \sin \theta \cos \left(\psi_c + \frac{\Delta \alpha}{2} \right), \tag{6.21}$$

$$F_{D2.f} = \frac{2V}{\lambda} \sin \theta \cos \left(\psi_c - \frac{\Delta \alpha}{2} \right), \tag{6.22}$$

$$F_{D1.a} = \frac{2V}{\lambda} \sin \theta \cos \left(180° - \psi_c - \frac{\Delta \alpha}{2} \right), \tag{6.23}$$

$$F_{D2.a} = \frac{2V}{\lambda} \sin \theta \cos \left(180° - \psi_c + \frac{\Delta \alpha}{2} \right). \tag{6.24}$$

At low speed of flight, the Doppler effect is not as considerable as at higher speed of flight, and from (6.21 to 6.24), we can see that a very narrow-band Doppler filter is required. When a technical difficulty arises with implementation of a very narrow-band filter, the wide-band Doppler filter also may be applied taking into account the azimuth angular size of a cell for measured NRCS with expression (3.24).

Let $\sigma°(U, \theta, \alpha + \psi_c, \Delta \alpha)$, $\sigma°(U, \theta, \alpha + 180° - \psi_c, \Delta \alpha)$, $\sigma°(U, \theta, \alpha + 180° + \psi_c, \Delta \alpha)$, and $\sigma°(U, \theta, \alpha - \psi_c, \Delta \alpha)$ be the NRCS corresponding to the selected identical cells (Fig. 6.4). Then, the average NRCS obtained with the fore-Doppler filter from the fore-cells $\sigma^{\circ}_{fd}(U, \theta, \alpha, \Delta\alpha, \psi_c)$ is (Nekrassov 2003)

$$
\begin{aligned}
\sigma^{\circ}_{fd}(U, \theta, \alpha, \Delta\alpha, \psi_c) &= \frac{1}{2}(\sigma^{\circ}(U, \theta, \alpha + \psi_c, \Delta\alpha \\
&\quad + \sigma^{\circ}(U, \theta, \alpha - \psi_c, \Delta\alpha)) \\
&= A(U, \theta) + k_1(\Delta\alpha)B(U, \theta)\cos\psi_c\cos\alpha \\
&\quad + k_2(\Delta\alpha)C(U, \theta)\cos(2\psi_c)\cos(2\alpha),
\end{aligned}
\tag{6.25}
$$

and the average NRCS obtained with the aft-Doppler filter from the aft-cells $\sigma^{\circ}_{ad}(U, \theta, \alpha, \Delta\alpha, \psi_c)$ is

$$
\begin{aligned}
\sigma^{\circ}_{ad}(U, \theta, \alpha, \Delta\alpha, \psi_c) &= \frac{1}{2}(\sigma^{\circ}(U, \theta, \alpha + 180^{\circ} - \psi_c, \Delta\alpha) \\
&\quad + \sigma^{\circ}(U, \theta, \alpha + 180^{\circ} + \psi_c, \Delta\alpha)) \\
&= A(U, \theta) - k_1(\Delta\alpha)B(U, \theta)\cos\psi_c\cos\alpha \\
&\quad + k_2(\Delta\alpha)C(U, \theta)\cos(2\psi_c)\cos(2\alpha).
\end{aligned}
\tag{6.26}
$$

Using (6.25) and (6.26), two possible wind directions $\psi_{w.1,2}$ can be found as follows

$$
\psi_{w.1,2} = \psi - \alpha_{1,2} \pm 180^{\circ},
\tag{6.27}
$$

where two possible up-wind directions are

$$
\alpha_{1,2} = \pm\arccos\left(\frac{\sigma^{\circ}_{fd}(U, \theta, \alpha, \Delta\alpha, \psi_c) - \sigma^{\circ}_{ad}(U, \theta, \alpha, \Delta\alpha, \psi_c)}{2k_1(\Delta\alpha)B(U, \theta)\cos\psi_c}\right).
\tag{6.28}
$$

Unfortunately, an ambiguity in the wind direction takes place in the measurement. Nevertheless, this ambiguity can be eliminated by another measurement after 45° change of the aircraft course. The nearest wind directions from pairs of wind directions measured before and after course change will give the true wind direction (Fig. 6.5).

The method for measuring sea-surface wind speed by the intensity of backscattered signal requires a minimization of the systematic error in the reflected signal magnitude. The use of a corner reflectors array for the external calibration of a scatterometer allows for decreasing this systematic error to approximately 1 dB (Melnik 1980). However, the external calibration with corner reflectors presents considerable difficulties at the operational measurement of sea wind.

Another way to decrease the systematic error in the magnitude of the reflected signal for a scatterometer with an inclined beam is to use an additional nadir-looking beam for calibration. The possible accuracy of such relative measuring cannot be much less than the accuracy of measurement with the corner reflectors (Melnik 1980). So, using (6.11) and (6.19) for such operational wind measurement, the following equation can be written

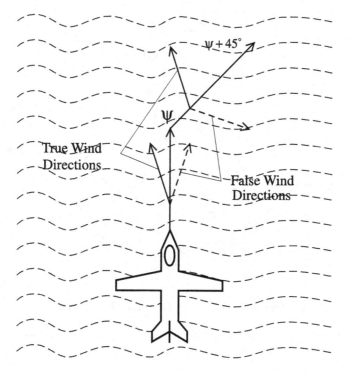

Fig. 6.5 Two-stage measurement scheme for the wind direction retrieval eliminating an ambiguity in the wind direction over water

$$A(U, \theta) \cdot 10^{-[G_1 + G_2 \log_{10}(U/0.93)]} = \frac{\overset{\circ}{\sigma}_{an}(U, \theta)}{\sigma^\circ(U, 0^\circ)}. \tag{6.29}$$

Then, using (6.20), the sea-surface wind speed can be found from (6.29)

$$U = \left(\frac{\overset{\circ}{\sigma}_{an}(U, \theta)}{0.93^{G_2} \cdot 10^{-G_1} a_0(\theta) \sigma^\circ(U, 0^\circ)} \right)^{1/[\gamma_0(\theta) - G_2]}. \tag{6.30}$$

Two possible up-wind directions obtained from the relative measurements can be defined from the following expression

$$\alpha_{1,2} = \pm \arccos \left(\frac{\overset{\circ}{\sigma}_{fd}(U, \theta, \alpha, \Delta\alpha, \psi_c) - \overset{\circ}{\sigma}_{ad}(U, \theta, \alpha, \Delta\alpha, \psi_c)}{2k_1(\Delta\alpha) B(U, \theta) \cos \psi_c} \times \frac{A(U, \theta)}{\overset{\circ}{\sigma}_{an}(U, \theta)} \right). \tag{6.31}$$

The wind direction ambiguity at relative measurement can be eliminated by the same way as at absolute measurement (Fig. 6.5).

As we can see, the algorithm for relative measuring is more preferable because it allows eliminating the influence of systematic error in the reflected signal magnitude in measurement result of sea-surface wind speed and direction.

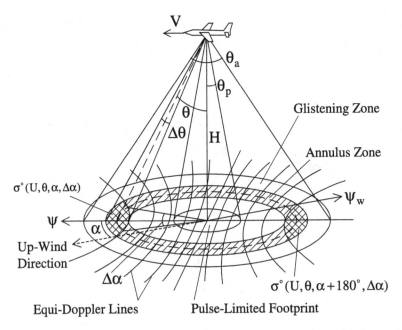

Fig. 6.6 Two-cell geometry for wind vector measurement with an antenna forming a *circle* footprint

A particular case takes place at a low speed of flight. In that case, the Doppler effect is not as considerable as at higher speed, and it is desirable to use NRCS data obtained from the cells for which the Doppler shift is the largest possible in current conditions. Hence, Doppler filtering should provide the azimuthal selection in the directions of 0 and 180° relative to the aircraft course as presented in Fig. 6.6.

Let $\sigma°(U, \theta, \alpha, \Delta\alpha)$ and $\sigma°(U, \theta, \alpha + 180°, \Delta\alpha)$ be the NRCS obtained with the fore-Doppler and aft-Doppler filters from the cells corresponding to the maximum value of the Doppler shift (Fig. 6.6). Then, the frequency limits for Doppler filters become (Nekrasov 2010b)

$$F_{D1.f} = -F_{D1.a} = \frac{2V}{\lambda} \sin\left(\theta - \frac{\Delta\theta}{2}\right), \tag{6.32}$$

$$F_{D2.f} = -F_{D2.a} = \frac{2V}{\lambda} \sin\left(\theta + \frac{\Delta\theta}{2}\right), \tag{6.33}$$

which, approximately, is $1/\cos\psi_c$ times higher in absolute value than in (6.18–6.21). Unfortunately, coarsest azimuth resolution resultant in that case is

$$\Delta\alpha = 2\arccos\left(\frac{\sin(\theta - \Delta\theta)}{\sin\theta}\right), \tag{6.34}$$

and NRCS model function (3.24), which considers the azimuth angular size of a cell, should be used.

In accordance with the geometry of Fig. 6.6, the wind vector can be estimated by means of the following algorithm.

The speed of wind can be found from (6.20) with absolute measurement, or from (6.30) with relative measurement. Two possible wind directions can be found from (6.27), taking into account that two possible up-wind directions with absolute measurement are as follows

$$\alpha_{1,2} = \pm \arccos \left(\frac{\sigma^\circ(U, \theta, \alpha, \Delta\alpha) - \sigma^\circ(U, \theta, \alpha + 180^\circ, \Delta\alpha)}{2k_1(\Delta\alpha)B(U, \theta)} \right), \tag{6.35}$$

and with relative measurement are

$$\alpha_{1,2} = \pm \arccos \left(\frac{\sigma^\circ(U, \theta, \alpha, \Delta\alpha) - \sigma^\circ(U, \theta, \alpha + 180^\circ, \Delta\alpha)}{2k_1(\Delta\alpha)B(U, \theta)} \times \frac{A(U, \theta)}{\sigma^\circ_{an}(U, \theta)} \right) \tag{6.36}$$

where $\sigma^\circ(U, \theta, \alpha, \Delta\alpha)$ is described by (3.24) and $\sigma^\circ(U, \theta, \alpha + 180^\circ, \Delta\alpha)$ is as follows

$$\sigma^\circ(U, \theta, \alpha + 180^\circ, \Delta\alpha) = A(U, \theta) + k_1(\Delta\alpha)B(U, \theta)\cos(\alpha + 180^\circ) \\ + k_2(\Delta\alpha)C(U, \theta)\cos(2(\alpha + 180^\circ)). \tag{6.37}$$

An ambiguity in the wind direction takes place in measurement based on the two-cell geometry as well. This ambiguity can be eliminated by the same way as with measurement based on the four-cell geometry via performing another measurement after 45° change of the aircraft course. The nearest wind directions from pairs of wind directions measured before and after course change will give the true wind direction (Fig. 6.5).

Now we can compare the measurement geometries. To evaluate the four-cell and two-cell geometries, an appropriate criterion should be found.

It is evident that the best geometry is such that allows carrying out wind measurement at a lower speed of flight and permits usage of the maximum possible difference between the up-wind and down-wind NRCS values. For algorithms based on the four-cell geometry (Fig. 6.4), the difference between the NRCS obtained with the fore-Doppler and aft-Doppler filters $\Delta\sigma^\circ_4(U, \theta, \alpha, \Delta\alpha, \psi_c)$ can be found from (6.25) to (6.26) as following (Nekrassov 2003)

$$\Delta\sigma^\circ_4(U, \theta, \alpha, \Delta\alpha, \psi_c) = 2k_1(\Delta\alpha)B(U, \theta)\cos\psi_c|\cos\alpha|, \tag{6.38}$$

and then, the maximum possible difference $\Delta\sigma^\circ_{4.\max}(U, \theta, \Delta\alpha, \psi_c)$ is

$$\Delta\sigma^\circ_{4.\max}(U, \theta, \Delta\alpha, \psi_c) = 2k_1(\Delta\alpha)B(U, \theta)\cos\psi_c. \tag{6.39}$$

At the same time, the two-cell geometry (Fig. 6.6) allowing another value for difference between the NRCS to be obtained with the fore-Doppler and aft-Doppler filters $\Delta\sigma^\circ_2(U, \theta, \alpha, \Delta\alpha)$, which, from (3.24) and (6.37), is

$$\Delta\overset{\circ}{\sigma}_2(U, \theta, \alpha, \Delta\alpha) = 2k_1(\Delta\alpha)B(U, \theta)|\cos\alpha|, \tag{6.40}$$

and maximum possible difference $\Delta\overset{\circ}{\sigma}_{2.\text{max}}(U, \theta, \Delta\alpha)$ in that case is

$$\Delta\overset{\circ}{\sigma}_{2.\text{max}}(U, \theta, \Delta\alpha) = 2k_1(\Delta\alpha)B(U, \theta). \tag{6.41}$$

Figure 3.4 and Table 3.2 show the dependence of coefficient $k_1(\Delta\alpha)$ on the provided azimuth resolution.

From (6.39) to (6.41), we can see that the two-cell geometry provides for the maximum possible difference between the measured up- and down-wind values of the NRCS, which is $1/\cos\psi_c$ times higher than is achievable with the four-cell geometry in spite of the fact that the two-cell geometry provides for worse azimuth resolution. The frequency limits of Doppler filters for the two-cell geometry are approximately $1/\cos\psi_c$ times higher in absolute value than for the four-cell geometry at the same speed of flight. This means that the geometry of Fig. 6.6 allows for carrying out wind measurement at a lower speed of flight, at least $1/\cos\psi_c$ times lower, than the geometry of Fig. 6.4 ($\sqrt{2}$ times lower relative to the four-cell geometry with $\psi_c = 45°$), and thus, the algorithms based on the two-cell geometry are more preferable for measuring the wind vector by such an airborne altimeter.

6.4 Wind Vector Estimation by an Airborne Radar Altimeter, Which has an Antenna with Modified Beam Shape

As the two-stage wind vector measurement by means of an ARA having an antenna with the same beamwidth in both vertical and horizontal planes is inconvenient for a pilot, especially under search-and-rescue missions or fire-fighting operations in coastal areas, another way to remove wind direction ambiguity should be found. It is evident that it can be achieved using an antenna with the modified beam shape that allows for obtaining NRCSs from additional azimuth directions.

If the antenna beam is wide enough, then the two annulus zones at incidence angles θ_1 and θ_2 can be formed as shown in Fig. 6.7. The zones have angular incidence widths $\Delta\theta_1$ and $\Delta\theta_2$, respectively (Nekrasov 2012b).

Let the altimeter antenna have different beamwidth in the vertical $\theta_{a.v}$ and horizontal $\theta_{a.h}$ planes ($\theta_{a.v} > \theta_{a.h}$) and form an egg-shaped footprint so that the longer axis of the footprint is at 45° to the right from the horizontal projection of the longitudinal axis of an aircraft as shown in Fig. 6.8.

Then, two annulus zones at incidence angles θ_1 and θ_2 ($\theta_1 < \theta_{a.h} < \theta_2 < \theta_{a.v}$) can be formed, and the range Doppler selection can provide the selection of cells with the NRCS $\sigma°(U, \theta_1, \alpha)$ and $\sigma°(U, \theta_1, \alpha + 180°)$ corresponding to the azimuth directions α and $\alpha + 180°$ from the first annulus, and the selection of a cell with the NRCS $\sigma°(U, \theta_2, \alpha + \psi_d)$ corresponding to the azimuth direction $\alpha + \psi_d$ from the

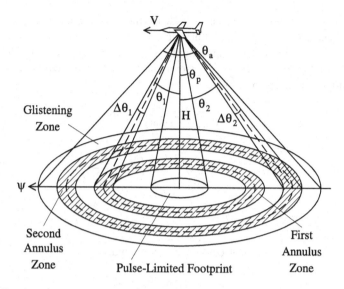

Fig. 6.7 Forming two annulus zones

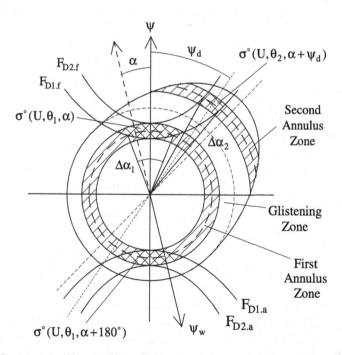

Fig. 6.8 Geometry for measuring the wind vector in the case of an antenna with different beamwidth in the vertical and horizontal planes forming the egg-shaped footprint when the longer axis of the footprint is at approximately 45° from the aircraft longitudinal axis

second annulus, where ψ_d is the angle between the aircraft course and the azimuth of the cell selected from the second annulus.

To provide the required azimuth angular sizes of cells of the first annulus $\Delta\alpha_1$ and the second annulus $\Delta\alpha_2$, as is shown in Fig. 6.8, the frequency limits for Doppler filters should be as follows

$$F_{D1.f} = -F_{D1.a} = \frac{2V}{\lambda}\sin\left(\theta_1 - \frac{\Delta\theta_1}{2}\right),\tag{6.42}$$

$$F_{D2.f} = -F_{D2.a} = \frac{2V}{\lambda}\sin\left(\theta_1 + \frac{\Delta\theta_1}{2}\right),\tag{6.43}$$

$$F_{D1.f} = -F_{D1.a} = \frac{2V}{\lambda}\sin\theta_2\cos\left(\psi_d + \frac{\Delta\alpha_2}{2}\right).\tag{6.44}$$

The azimuth angular size of cells of the first annulus is

$$\Delta\alpha_1 = 2\arccos\left(\frac{\sin(\theta_1 - \Delta\theta_1)}{\sin\theta_1}\right).\tag{6.45}$$

From (6.42 to 6.44), the azimuth location of a cell of the second annulus ψ_d and its angular size in the horizontal plane are as follows

$$\psi_d = 0.5\left[\arccos\left(\frac{\sin(\theta_1 - 0.5\Delta\theta_1)}{\sin\theta_2}\right) + \arccos\left(\frac{\sin(\theta_1 + 0.5\Delta\theta_1)}{\sin\theta_2}\right)\right],\tag{6.46}$$

$$\Delta\alpha_2 = \arccos\left(\frac{\sin(\theta_1 - 0.5\Delta\theta_1)}{\sin\theta_2}\right) - \arccos\left(\frac{\sin(\theta_1 + 0.5\Delta\theta_1)}{\sin\theta_2}\right).\tag{6.47}$$

The wind speed can be found from the following equation

$$U = \left(\frac{A(U,\theta_1)}{a_0(\theta_1)}\right)^{1/\gamma_0(\theta_1)} = \left(\frac{\overset{\circ}{\sigma}_{an}(U,\theta_1)}{a_0(\theta_1)}\right)^{1/\gamma_0(\theta_1)}.\tag{6.48}$$

Two possible up-wind directions $\alpha_{1an.1,2}$ can be found from the NRCS values obtained from cells of the first annulus

$$\alpha_{1an.1,2} = \pm\arccos\left(\frac{\sigma°(U,\theta_1,\alpha) - \sigma°(U,\theta_1,\alpha + 180°)}{2k_1(\Delta\alpha_1)B(U,\theta_1)}\right).\tag{6.49}$$

Then, substituting these possible up-wind directions into (3.24), two calculated NRCS values for the second annulus $\overset{\circ}{\sigma}_{cal}(U,\theta_2,\alpha_1 + \psi_d)$ and $\overset{\circ}{\sigma}_{cal}(U,\theta_2,\alpha_2 + \psi_d)$ are obtained. One NRCS value from calculated NRCS values, which is nearest to the NRCS value $\sigma°(U,\theta_2,\alpha + \psi_d)$ measured from the second annulus, will

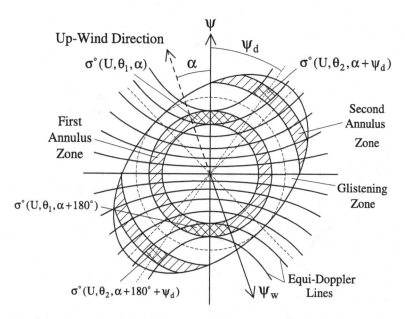

Fig. 6.9 Geometry for measuring the wind vector in the case of an antenna with different beamwidth in the *vertical* and *horizontal* planes forming the *ellipse* footprint when the longer axis of the footprint is at approximately 45° from the aircraft longitudinal axis

correspond to the true up-wind direction α (one from $\alpha_{1an.1}$ and $\alpha_{1an.2}$), and then, wind direction can be found from (4.20).

Now, let the altimeter antenna have different beamwidth in the vertical and horizontal planes ($\theta_{a.v} > \theta_{a.h}$) forming an ellipse footprint so as the longer axis of the footprint is at 45° to the right from the horizontal projection of the longitudinal axis of an aircraft as shown in Fig. 6.9.

Then, two annulus zones at incidence angles θ_1 and θ_2 ($\theta_1 < \theta_{a.h} < \theta_2 < \theta_{a.v}$) could be formed, and the range Doppler selection can facilitate identification of cells with NRCS $\sigma°(U, \theta_1, \alpha)$ and $\sigma°(U, \theta_1, \alpha + 180°)$ corresponding to the azimuth directions α and $\alpha + 180°$ from the first annulus, and identification of cells with NRCS $\sigma°(U, \theta_2, \alpha + \psi_d)$ and $\sigma°(U, \theta_2, \alpha + 180° + \psi_d)$ corresponding to the azimuth directions $\alpha + \psi_d$ and $\alpha + 180° + \psi_d$ from the second annulus.

To provide the required azimuth angular sizes of cells of the first and second annuluses as shown in Fig. 6.10, frequency limits for Doppler filters should be as in (6.42–6.44) and (Nekrasov 2008a)

$$F_{D2.f} = -F_{D2.a} = \frac{2V}{\lambda} \sin \theta_2 \cos \left(\psi_d - \frac{\Delta \alpha_2}{2} \right). \qquad (6.50)$$

The azimuth angular size of the first annulus cells is given by (6.45). The azimuth locations of the second annulus cells ψ_d, $180° + \psi_d$, and their angular size in the horizontal plane are described by (6.46) and (6.47). The speed of wind can be found from (6.48).

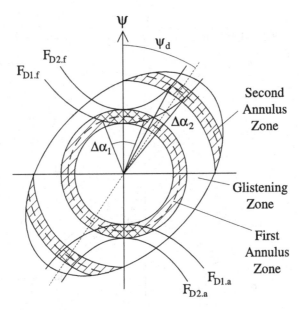

Fig. 6.10 Determing selected cells and their angular sizes in *horizontal* plane in a case of the antenna with the different beamwidth in the *vertical* and *horizontal* planes forming the *ellipse* footprint when the longer axis of the footprint at approximately 45° from the aircraft longitudinal axis

Two possible up-wind directions $\alpha_{1an.1,2}$ can be found from NRCS values obtained from the cells of the first annulus using (6.49), and another two possible up-wind directions $\alpha_{2an.1,2}$ can be found from the NRCS values obtained from the cells of the second annulus using the following equation (Nekrasov 2007)

$$\alpha_{2an.1,2} = \pm \arccos \left(\frac{\sigma°(U, \theta_2, \alpha + \psi_d) - \sigma°(U, \theta_2, \alpha + 180° + \psi_d)}{2k_1(\Delta\alpha_2)B(U, \theta_2)} \right). \quad (6.51)$$

The nearest up-wind directions of pairs of the up-wind directions obtained (one from $\alpha_{1an.1,2}$ and one from $\alpha_{2an.1,2}$) will give the true up-wind direction α, and then, the navigational direction of wind can be found from (4.20).

From (6.46), we can see that the azimuth locations of cells of the second annulus do not always coincide with the longer axis of the ellipse (egg-shaped) footprint at 45° from the longitudinal axis of an aircraft, because they depend on the incidence angles corresponding to the first and the second annulus zones and angular incidence width of the first annulus. Therefore, the antenna should be installed so as the longer axis of the ellipse (egg-shaped) footprint coincides with the azimuth locations of cells of the second annulus in operating mode.

The ARA wind measurement is started when a stable rectilinear flight at the given altitude and speed of flight has been established. The measurement is finished when a required number of NRCS samples for each cell or cell pair is obtained.

6.5   Conclusions to Water-Surface Wind Retrieval Using Airborne Radar Altimeter

The study has shown that wind vector over sea can be measured by means of the ARA employed as a nadir-looking wide-beam short-pulse scatterometer in conjunction with Doppler filtering. The measuring instrument should be equipped with two additional Doppler filters (a fore-Doppler filter and an aft-Doppler filter) to provide for spatial selection under the wind measurements.

Such an altimeter should operate at a Ku- or X-band (or at least at a C-band) using a horizontal transmit and receive polarizations. Shorter radar wavelength provides Doppler selection at a lower speed of flight, and at the Ku-band, the up-wind–down-wind and up-wind–cross-wind differences in NRCS values at medium incidence angles (for the wind speed of 3–24 m/s) are greater than at the lower bands. Horizontal transmit and receive polarizations provide greater up-wind–down-wind differences in NRCS values at medium incidence angles than the vertical transmit and receive polarizations.

The two-cell geometry for retrieval of the wind vector, when the antenna forms a circle footprint and the spatially selected cells are located at the directions of $0°$ and $180°$ relative to the aircraft course, is more preferable than the four-cell geometry. The two-cell geometry provides the frequency limits of the Doppler filters and the maximum possible difference between the measured up- and down-wind values of the NRCS, which are $1/\cos\psi_c$ times higher than it is possible with the four-cell geometry. It means that the two-cell geometry allows carrying out wind measurements at a lower speed of flight than the four-cell geometry, at least $1/\cos\psi_c$ times lower ($\sqrt{2}$ times lower relative to the four-cell geometry with $\psi_c = 45°$).

Unfortunately, ambiguity in wind direction appears in measurements when the antenna forming a circle footprint is used. Nevertheless, this ambiguity can be eliminated by another measurement after $45°$ change of the aircraft course when the nearest wind directions from pairs of wind directions measured before and after course changing will give the true wind direction. Otherwise, to avoid such an inconvenience for a pilot, especially under search and rescue missions or fire-fighting operations in coastal areas, the antenna with the modified beam shape can be used.

The incidence angle for the first annulus zone should be no less than $20°$, but the incidence angle for the second annulus zone should tend to $45°$ when an antenna with the modified beam shape is used. The antenna should have different beam-widths in the vertical and horizontal planes ($\theta_{a.v} > \theta_{a.h}$) and form the ellipse footprint, or at least the egg-shaped footprint, so that the longer axis of the footprint is at approximately $45°$ from the horizontal projection of the longitudinal axis of an aircraft. It is desirable that the antenna should be installed so as the longer axis of the ellipse (egg-shaped) footprint coincides with the azimuth locations of cells of the second annulus in operating mode.

For wind measurement, an antenna with the ellipse footprint is more preferable compared to an antenna forming the egg-shaped footprint. The ellipse footprint allows obtaining the NRCS values from four significantly different azimuth directions that provides better wind vector estimation than it can be achieved with NRCSs obtained only from three azimuth directions when the egg-shaped footprint is used.

As the method for measuring sea-surface wind by the intensity of backscattered signal requires minimization of the systematic error in the reflected signal magnitude, and as the application of an array of corner reflectors for external calibration of the instrument has considerable difficulties during operational measurement of sea wind, relative measurement using additional NRCS data obtained from nadir also can be used for the calibration. Therefore, the proposed algorithms based on the relative measurement are more preferable for such estimation of the wind vector.

References

Gommenginger C, Martin-Puig C, Dinardo S, Cotton D, Benveniste J (2012) Improved altimetric performance of Cryosat-2 SAR mode over the open ocean and the coastal zone. EGU General Assembly 2012. vol 14, Vienna, Austria, 22–27 Apr 2012, Geophysical Research Abstracts, p 6580

Hammond DL, Mennella RA, Walsh EJ (1977) Short pulse radar used to measure sea surface wind speed and SWH. IEEE Trans Antennas Propag 25(1):61–67

Jackson FC, Walton WT, Hines DE, Walter BA, Peng CY (1992) Sea surface mean square slope from Ku-band backscatter data. J Geophys Res 97((C7)):11411–11427

Kayton M, Fried WR (1997) Avionics navigation systems. Wiley, New York, p 773

Komjathy A, Zavorotny V, Axelrad P, Born G, Garrison JL (2000) GPS signal scattering from sea surface: wind speed retrieval using experimental data and theoretical model. Remote Sens Env 73(2):162–174

Komjathy A, Armatys M, Masters D, Axelrad P, Zavorotny VU, Katzberg SJ (2001) Developments in using GPS for oceanographic remote sensing: retrieval of ocean surface wind speed and wind direction. In: Proceedings of the ION National technical meeting, Long Beach, CA, USA, 22–24 Jan 2001, p 9

Martin-Puig C, Ruffini G (2009) SAR altimeter retracker performance bound over water surface. In: Proceedings of IGARSS 2009. vol 5, Cape Town, South Africa, 12–17 Jul 2009, pp 449–452

Melnik YuA (1980) Radar methods of the Earth exploration. Sovetskoye Radio, Moscow, USSR, p 264, in Russian

Nekrasov A (2007) Measurement of the wind vector over sea by an airborne radar altimeter, which has an antenna with the ellipse beam shape. In: Proceedings of APMC 2007, Bangkok, Thailand, 11–14 Dec 2007, pp 91–94

Nekrasov A (2008) Measurement of the wind vector over sea by an airborne radar altimeter having an antenna with the different beamwidth in the vertical and horizontal planes. IEEE Geosci Remote Sens Lett 5(1):31–33

Nekrasov A (2010b) Microwave measurement of the wind vector over sea by airborne radars. In: Mukherjee M (ed) Advanced microwave and millimeter wave technologies: semiconductor devices, circuits and systems, In-Tech, Vukovar, pp 521–548

Nekrasov A (2012b) Measurement of the wind vector over the water surface by an airborne radar altimeter, which has an antenna forming the egg-shaped footprint. Radar Science and Technology 10(5):460–466

Nekrassov A (2001) Measurement of the sea surface wind speed and direction by an airborne microwave radar altimeter. GKSS Report No. GKSS/2001/38, Geesthacht, Germany, p 17

Nekrassov A (2002) On airborne measurement of the sea surface wind vector by a scatterometer (altimeter) with a nadir-looking wide-beam antenna. IEEE Trans Geosci Remote Sens 40 (10):2111–2116

Nekrassov A (2003) Airborne measurement of the sea surface wind vector by a microwave radar altimeter at low speed of flight. IEICE Trans Electron E86-C(8):1572–1579

Raney RK (1998) The delay/Doppler radar altimeter. IEEE Trans Geosci Remote Sens 36 (5):1578–1588

Spencer WM, Tsai WY, Long DG (2000a) High resolution scatterometry by simultaneous range/ Doppler discrimination. In: Proceedings of IGARSS 2000, Honolulu, Hawaii, USA, 24–28 Jul 2000, pp 3166–3168

Spencer WM, Wu C, Long DG (2000b) Improved resolution backscatter measurements with the SeaWinds pencil-beam scatterometer. IEEE Trans Geosci Remote Sens 38(1):89–104

Wang Q, Gogineni S (1991) A numerical procedure for recovering scattering coefficients from measurements with wide-beam antennas. IEEE Trans Geosci Remote Sens 29(5):778–783

Chapter 7
Near-Nadir Wind Estimation Over Water with Airborne Precipitation Radar

7.1 Airborne Precipitation Radar-2 Geometry

Remote sensing of sea-surface wind by means of a radar altimeter is based on specular returns from the water surface. A wind retrieval algorithm deriving the wind speed from the NRCS at nadir incidence angle is used in that case (Zhao and Toba 2003; Zieger 2010). Unfortunately, for some special cases of measurements, specific radar instrument configuration or its installation on board of an aircraft, the near-nadir NRCSs only may be available during the measurement. Such a case takes place for example when a measuring geometry of the Airborne Precipitation Radar-2 (APR-2) is used (Sadowy et al. 2003).

APR-2 is an airborne, dual-frequency (Ku- and Ka-band), dual-polarization Doppler rain profiling radar. Its antenna scans in the $\pm 25°$ cross-track elevation range allowing the radar to measure atmospheric precipitation and sea-surface NRCS. Each cross-track scan is sampled at 23 positions (since January 2003 the 24th position is used for noise floor measurements) (Tanelli et al. 2004, 2006). Since during the Wakasa Bay Experiment in 2003, the radar was configured to operate at a forward-looking angle of $3°$ and aircraft pitch angle was about $+1°$, we may assume that the incidence angle θ_0 in that case was no more than $4°$.

Let a flying apparatus equipped with a radar similar to APR-2 perform a horizontal rectilinear flight with speed V at some altitude H above mean sea surface, the radar operate in a scatterometer mode, the radar antenna have beamwidths in the vertical plane $\theta_{a.v}$ and in the horizontal plane $\theta_{a.h}$, scan periodically in the given $\pm \theta_s$ cross-track elevation range, and provide NRCS measuring from a current selected cell as shown in Fig. 7.1. Then, measured NRCS will correspond to the current near-surface wind speed U, incidence angle θ, and azimuth illumination angle α relative to the up-wind direction.

© The Author(s), under exclusive license to Springer Nature Switzerland AG 2021
A. Nekrasov, *Foundations for Innovative Application of Airborne Radars*,
SpringerBriefs in Earth Sciences, https://doi.org/10.1007/978-3-030-62942-7_7

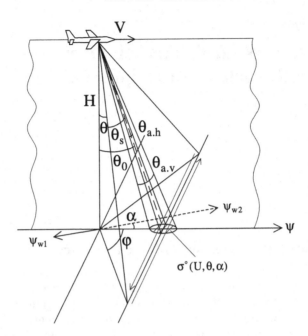

Fig. 7.1 Airborne precipitation radar measuring geometry

In accordance with the measurement geometry of Fig. 7.1, the current incidence angle and horizontal angle φ of the selected sell relative to the aircraft course ψ are as follows

$$\theta = \arccos(\cos\theta_0 \, \cos\theta_s), \tag{7.1}$$

$$\varphi = \arccos\left(\frac{\tan\theta_s}{\sin\theta_0}\right) \quad for \quad \theta_0 > 0°, \tag{7.2}$$

where θ_0 is the incidence angle for the selected cell when scanning beam direction in the vertical plane coincides with the aircraft course,

$$\theta_0 = \theta_m + \theta_{fl}, \tag{7.3}$$

where θ_m is the antenna mounting angle and θ_{fl} is the aircraft pitch angle.

Thus, the measuring geometry allows obtaining a number of near-nadir NRCS values for various incidence and azimuth illumination angles.

7.2 Near-Nadir Wind Retrieval with Airborne Precipitation Radar Geometry

Let no precipitation be over the water surface at the wind measurement. When near-nadir NRCSs are available for the incidence angles of up to $15° − 20°$ during measurement, the water surface is considered anisotropic and described by Gaussian statistics, and the incidence angle dependence for the quasi-specular scattering is expressed as follows (Donelan and Pierson 1987)

$$\sigma°(\theta, \alpha) = \frac{|R(0°)|^2}{2S_u S_c} \cdot \frac{1}{\cos^4 \theta} \exp\left(-\frac{\tan^2 \theta}{2S^2(\alpha)}\right),\tag{7.4}$$

where S_u and S_c are the up-wind and cross-wind standard deviations of slopes, and $S^2(\alpha)$ is water-surface slope variance in azimuthal direction α (Donelan and Pierson 1987),

$$S^2(\alpha) = \frac{S_u^2 S_c^2}{S_u^2 \sin^2 \alpha + S_c^2 \cos^2 \alpha}.\tag{7.5}$$

As three parameters are unknown in (7.4) and azimuth water-surface slope variance is represented in the elliptical form of (7.5), usually at least five NRCSs from different azimuth angles are required to calculate them.

Let a number of five or more NRCSs be available in accordance with the measurement geometry of Fig. 7.1. Then, taking into account the measuring geometry and using (7.4) and (7.5), the following system of equations for those NRCSs could be written (Nekrasov et al. 2018)

$$\left\{ \begin{array}{l} \sigma^\circ(\theta_1, \alpha) = \dfrac{|R(0^\circ)|^2}{2 S_u S_c} \cdot \dfrac{1}{\cos^4 \theta_1} \exp\left(-\dfrac{\tan^2 \theta_1}{\dfrac{2 S_u^2 S_c^2}{S_u^2 \sin^2 \alpha + S_c^2 \cos^2 \alpha}} \right), \\[2em]

\sigma^\circ(\theta_2, \alpha + \varphi_2) = \dfrac{|R(0^\circ)|^2}{2 S_u S_c} \cdot \dfrac{1}{\cos^4 \theta_2} \\[1.5em]

\times \exp\left(-\dfrac{\tan^2 \theta_2}{\dfrac{2 S_u^2 S_c^2}{S_u^2 \sin^2(\alpha + \varphi_2) + S_c^2 \cos^2(\alpha + \varphi_2)}} \right), \\[2em]

\sigma^\circ(\theta_2, \alpha - \varphi_2) = \dfrac{|R(0^\circ)|^2}{2 S_u S_c} \cdot \dfrac{1}{\cos^4 \theta_2} \\[1.5em]

\times \exp\left(-\dfrac{\tan^2 \theta_2}{\dfrac{2 S_u^2 S_c^2}{S_u^2 \sin^2(\alpha - \varphi_2) + S_c^2 \cos^2(\alpha - \varphi_2)}} \right), \\[1em]

\cdots\cdots\cdots\cdots\cdots\cdots\cdots\cdots\cdots\cdots\cdots\cdots \\[1em]

\sigma^\circ(\theta_N, \alpha + \varphi_N) = \dfrac{|R(0^\circ)|^2}{2 S_u S_c} \cdot \dfrac{1}{\cos^4 \theta_N} \\[1.5em]

\times \exp\left(-\dfrac{\tan^2 \theta_N}{\dfrac{2 S_u^2 S_c^2}{S_u^2 \sin^2(\alpha + \varphi_N) + S_c^2 \cos^2(\alpha + \varphi_N)}} \right), \\[2em]

\sigma^\circ(\theta_N, \alpha - \varphi_N) = \dfrac{|R(0^\circ)|^2}{2 S_u S_c} \cdot \dfrac{1}{\cos^4 \theta_N} \\[1.5em]

\times \exp\left(-\dfrac{\tan^2 \theta_N}{\dfrac{2 S_u^2 S_c^2}{S_u^2 \sin^2(\alpha - \varphi_N) + S_c^2 \cos^2(\alpha - \varphi_N)}} \right), \end{array} \right. \tag{7.6}$$

where $\sigma^\circ(\theta_1, \alpha)$, $\sigma^\circ(\theta_N, \alpha + \varphi_N)$, $\sigma^\circ(\theta_N, \alpha - \varphi_N)$ are NRCSs corresponding to the appropriate incidence angle $\theta_1, \ldots, \theta_N$; $\varphi_2, \ldots, \varphi_N$ are the azimuth angles between the azimuth angle $\varphi_1 = 0^\circ$ corresponding to the scanning beam direction in the vertical plane, which coincides with the aircraft course (and so φ_1 has been eliminated from the first equation of the system).

Solving the system of Eq. (7.6) approximately using searching procedure within the ranges of discrete values of possible solutions, the up-wind direction and the up-wind and cross-wind standard deviations of slopes can be found. Then, as the azimuth water-surface slope variance has been assumed to be elliptical (7.5), the wind direction can be found but with an ambiguity of 180° (Nekrasov et al. 2013)

$$\psi_{w1,2} = \begin{cases} \alpha \pm 180°, \\ \alpha. \end{cases} \tag{7.7}$$

Meanwhile, to estimate the wind speed over water, the relationship between the up-wind and cross-wind water-surface slope variances and wind speed for an appropriate frequency band similar to Cox and Munk equations may be used (Cox and Munk 1954)

$$S_u^2 = 0.00316U_{12.5}, \tag{7.8}$$

$$S_c^2 = 0.003 + 0.00192U_{12.5}, \tag{7.9}$$

where $U_{12.5}$ is the wind speed at 12.5-meter height.

Specifically, wind speed could be calculated from multiplication of such equations for up-wind and cross-wind water-surface slope variances. For a Cox and Munk case, the wind speed is calculated from the following equation

$$S_u^2 S_c^2 = 0.00316U_{12.5} \cdot (0.003 + 0.00192U_{12.5}) \tag{7.10}$$

obtained from (7.8) and (7.9).

The wind measurement using the measurement geometry of Fig. 7.1 is started when a stable rectilinear flight at the given altitude and speed of flight has been established. The measurement is finished when a required number of NRCS samples for each given incidence angle is obtained. To obtain a greater number of NRCS samples for each incidence angle, several consecutive beam sweeps may be used.

Table 7.1 Values assumed for the current incidence and horizontal angles

N	θ, deg.	φ, deg.
1	4	0
2	6	48
3	8	60
4	10	67
5	12	71
6	14	74
7	16	76
8	18	77
9	20	79
10	22	80

The maximum altitude of the method applicability depends on the APR-2 measuring geometry. Assuming the waves and winds conditions are identical in the observation area which length and width do not exceed 15–20 km, the maximum altitude is about 25 km for the wind and wave measurement with APR-2 at the rectilinear flight.

7.3   Simulation of Wind Vector Retrieval Based on Airborne Precipitation Radar Geometry

To verify the feasibility of the measuring geometry of Fig. 7.1 and the algorithm proposed, a simulation of the wind vector retrieval has been performed.

To simplify calculations, the values for current incidence and horizontal angles have been assumed in accordance with Table 7.1 to be used in the simulation. The "measured" NRCS values were generated using Gaussian distribution and (7.4) along with the Cox and Munk up-wind and cross-wind water-surface slope variances (7.8) and (7.9) adapted to the wind speed at the 10-meter height using (6.10)

$$S_u^2 = 0.00306965U, \tag{7.11}$$

$$S_c^2 = 0.003 + 0.0018651U. \tag{7.12}$$

Thus, wind speed has been calculated from the following equation

$$S_u^2 S_c^2 = 0.00306965U \cdot (0.003 + 0.0018651U). \tag{7.13}$$

Then, the proposed algorithm for recovering wind speed and direction has been tested using a Monte Carlo method with 50 trials at the wind speeds of 4–8 m/s under various numbers of equations in the system (7.6) and averaged NRCS samples for each selected cell. Simulation results are given in azimuth range of $0° - 179°$ due to the elliptical form of the azimuth water-surface slope variance assumed.

Figure 7.2 represents simulation results using 19 equations ($N = 10$) in the system of Eq. (7.6) for incidence angles of $4° - 22°$ with 1,565 averaged NRCS samples for each selected cell. The maximum error of wind speed retrieval in this case is only 0.8 m/s, and the maximum error of wind direction estimation is $17°$.

The results for 13 equations ($N = 7$) in the system of Eq. (7.6) for incidence angles of $4° - 16°$ with 1,565 averaged NRCS samples for each selected cell are shown in Fig. 7.3. The maximum error of wind retrieval for wind speed is 0.9 m/s and for wind direction is $18°$.

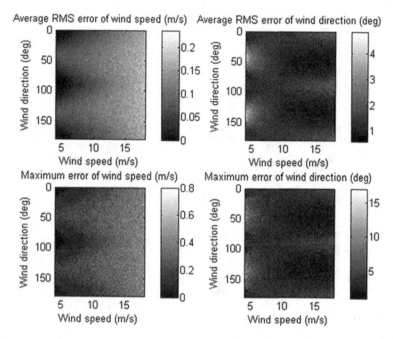

Fig. 7.2 Simulation results for 19 equations ($N = 10$) in the system of Eq. (7.6) for incidence angles of $4° - 22°$ with 1,565 averaged NRCS samples for each selected cell

The maximum wind speed error of 1.3 m/s and the maximum wind direction error of 32° take place for the same number of equations and incidence angles range at a decreased number of 785 averaged NRCS samples for each selected cell that is depicted in Fig. 7.4.

Figure 7.5 shows clearly the disadvantage of using 3 equations ($N = 2$) in the system of Eq. (7.6) for incidence angles of $4° - 6°$ with 1,565 averaged NRCS samples for each selected cell for the wind direction retrieval, as the maximum error of the wind direction estimation is 90° that was predictable based on Sect. 7.2 (at least five NRCSs from different azimuth angles are required) and due to low anisotropy of azimuth backscatter at those incidence angles. Nevertheless, a three equation case is still quite applicable to wind speed measurement, as it provides for the maximum wind speed error of 1.2 m/s only.

These examples of simulation results show us clearly the feasibility of APR measuring geometry of Fig. 7.1 and the algorithm proposed for wind vector estimation over the water surface. The "measured" wind speed is within a typical accuracy of ±2 m/s of wind speed measurement by a scatterometer. The "measured" wind directions are also within a typical accuracy of ±20° for a scatterometer in cases of 1565 averaged NRCS samples for each selected cell within of ranges of incidence angles of $4° - 22°$ and of $4° - 16°$.

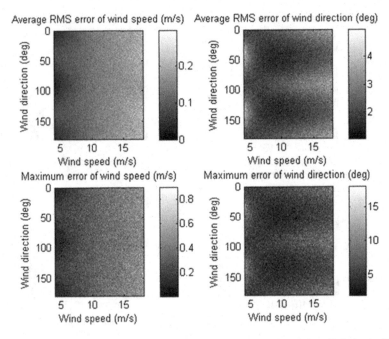

Fig. 7.3 Simulation results for 13 equations $(N = 7)$ in the system of Eq. (7.6) for incidence angles of $4° - 16°$ with 1,565 averaged NRCS samples for each selected cell

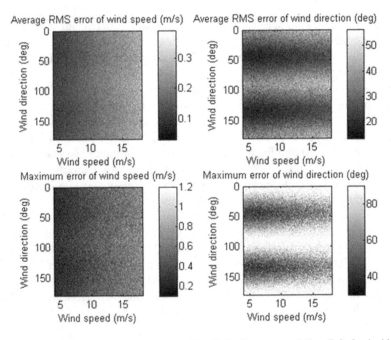

Fig. 7.4 Simulation results for 13 equations $(N = 7)$ in the system of Eq. (7.6) for incidence angles of $4° - 16°$ with 785 averaged NRCS samples for each selected cell

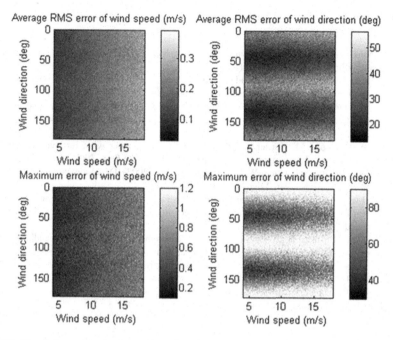

Fig. 7.5 Simulation results for 3 equations ($N = 2$) in the system of Eq. (7.6) for incidence angles of $4° - 6°$ with 1,565 averaged NRCS samples for each selected cell

7.4   Conclusions to Near-Nadir Wind Estimation Over Water with Airborne Precipitation Radar

The study has shown that airborne radar instrument for measurement geometry similar to the APR-2 geometry and operating in a scatterometer mode can be applied to remote measurement of sea-surface wind speed and direction at near-nadir incidence angles based on the measuring algorithm developed in case of precipitation absence. Otherwise, precipitations should also be taken into account in the wind measurement algorithm. As azimuth water-surface slope variance has been assumed to be elliptical, wind direction can be estimated, unfortunately, only with an ambiguity of 180°. It means that in principle the APR-2 also can be applied to wind retrieval over water surface during a rectilinear flight in addition to its typical meteorological application.

References

Cox C, Munk WH (1954) Measurement of the roughness of the sea surface from photographs of the Sun's glitter. J Opt Sot Am 44(11):838–850

Donelan MA, Pierson WJ (1987) Radar scattering and equilibrium ranges in wind-generated waves with application to scatterometry. J Geophys Res 93(C5):4871–5029

Nekrasov A, Ouellette JD, Majurec N, Johnson JT (2013) A study of sea surface wind vector estimation from near nadiral cross-track-scanned backscatter data. IEEE Geosci Remote Sens Lett 10(8):1503–1506

Nekrasov A, Gamcová M, Kurdel P, Labun J (2018) On off-nadir wind retrieval over the sea surface using APR-2 or similar radar geometry. Int J Remote Sens 39(18):5934–5942

Sadowy GA, Berkun AC, Chun W, Im E, Durden SL (2003) Development of an advanced airborne precipitation radar. Microwave J 46(1):84–98

Tanelli S, Meagher JP, Durden SL, Im E (2004) Processing of high resolution, multiparametric radar data for the Airborne Dual-Frequency Precipitation Radar APR-2. Proc SPIE 5654:25–32. https://doi.org/10.1117/12.579015

Tanelli S, Durden SL, Im E (2006) Simultaneous measurements of Ku- and Ka-band sea surface cross sections by an airborne radar. IEEE Geosci Remote Sens Lett 3(3):359–363

Zhao D, Toba Y (2003) A spectral approach for determining altimeter wind speed model functions. J Oceanogr 59(2):235–244

Zieger S (2010) Long term trends in ocean wind speed and wave height. Thesis, Swinburne University of Technology, Melbourne, Australia, p 177

Conclusions

This monograph demonstrates that an airborne FM-CW demonstrator system, Doppler navigation system, airborne weather radar, airborne radar altimeter, and airborne precipitation radar functionalities can be extended to operational oceanography, as well as to meteorology and navigation. For these purposes, all radars considered should be enhanced so as to operate in a scatterometer mode.

To perform measurement of water-surface backscattering signature, an aircraft equipped with the airborne FM-CW demonstrator system, Doppler navigation system, or airborne weather radar should perform a horizontal circle flight. The wind vector can be recovered well from obtained NRCS azimuth curves.

Wind vector can also be measured with a Doppler navigation system, airborne weather radar, airborne radar altimeter, and airborne precipitation radar during a horizontal rectilinear flight of an aircraft. Doppler navigation system having the three- or four-beam fixed or roll-and-pitch-stabilized antenna system should be used as a multi-beam scatterometer. Airborne weather radar should be employed in the ground-mapping mode as a scatterometer scanning periodically through an azimuth in narrow, medium or wide sector. Airborne radar altimeter should operate as a nadir-looking wide-beam short-pulse scatterometer in conjunction with Doppler filtering. Airborne precipitation radar should be used as a scatterometer when precipitation is absent, or its influence should be taken into account.

The principles considered and algorithms proposed in this monograph can be used for enhancement of those radars, for designing an airborne radar system for operational measurement of sea roughness characteristics, and for estimation of wind speed and direction over water. They may also be used for ensuring safe landing of amphibious aircraft on the water surface, for example under search-and-rescue missions or fire fighting in coastal areas and fire risk regions.

© The Author(s), under exclusive license to Springer Nature Switzerland AG 2021   113
A. Nekrasov, *Foundations for Innovative Application of Airborne Radars*,
SpringerBriefs in Earth Sciences, https://doi.org/10.1007/978-3-030-62942-7

Printed in the United States
by Baker & Taylor Publisher Services